SVEN BERTELSEN

THE UNRULY PROJECT

A new understanding
of its nature and
management

SVEN BERTELSEN aps

THE UNRULY PROJECT
© Sven Bertelsen 2015, 2017

The Danish version is published in 2015 by Lean Construction – DK, supported by Realdania.
The English version is published by Sven Bertelsen aps.
Danish editor: Poul Høegh Østergaard
English editor: Glenn Ballard
Cover and layout: Claus Lynggaard
Typeset: Eames Century and House Gothic
1st edition 2015 (in Danish); 2017 (in English)
ISBN 978-87-993283-1-4

Contents

Homepage: www.theunrulyproject.com

The editor's foreword

Sven Bertelsen and I have been close friends since 1999 when he began participating in the International Group for Lean Construction. His contributions to knowledge and practice in the construction industry began much earlier in his career with the Danish engineering consultancy firm NIRAS. Sven often enriches his conversation and teaching with stories from his NIRAS projects in Greenland, where the barriers of distance inspired a culture of inquiry, problem solving and deep respect for individuals. Sven was a driver of experimentation in production control and logistics in the early 1990s initiative Byggelogistik *(Construction Logistics)*, and has been a guiding hand in Lean Construction-DK, the first international affiliate of the Lean Construction Institute. His influence extends well beyond Denmark to the worldwide construction industry. He is truly a 'Grand Old Man' whose essays have arisen from his love affair with projects. This Foreword is a guide to reading and learning from his reflections and rich experience.

Some things about Sven to keep in mind when reading his essays:
Sven creates ideas – ways of seeing the world that reveal new opportunities and needs. Not a person who gets into

the details, he hopes to inspire others to bring his ideas into practice.

Sven is a practitioner who reflects on his experience. Not a researcher, he tends to start from what he's seen and done rather than from the experience of others.

And yet, Sven's idea creation has been informed by some key concepts and theories from others; including the theory of complex adaptive systems, the transformation-flow-value (TFV) theory of production, the Last Planner system of production planning and control, and the concept of bottlenecks.

What are Sven's main ideas?
First and foremost he believes projects are complex and that we tend to seriously underestimate just how complex they are. Complexity cannot be managed as described in the project management textbooks, but rather in the way an unruly horse is ridden. The ride is always full of twists and turns, so the rider has to be alert and adaptive. The rider must expect the horse to be unruly, and not make the mistake of thinking that this time out of the stable, everything will go smoothly all by itself.

Managing projects necessarily involves managing the flows of information, materials, labor, equipment – everything needed to build constructed objects. Yet flow management has been neglected. This is the more problematic because the potential for turbulence in flows is ill understood, making it more likely, to mix metaphors, that the horse will throw the rider.

Along with advocating attention to bottlenecks, one of Sven's ideas is that we explore the applicability of concepts and theories from hydraulics to deepen our understanding of flows in project production systems.

Despite the fact that projects come into being in order to deliver something valued by the buyers, the production of value is also much neglected. Sven suggests having one manager for value, one for flow, and one for operations. But he stresses the need for coordinating and balancing; knowing which dial to turn in order to achieve project objectives.

Don't be discouraged

Sven starts the book with a salvo at project management thinking and practice. To paraphrase: 'We don't understand at all what it is we project managers are trying to manage.' Hearing this, the reader might be inclined to think managing projects is futile and begin searching for new employment. But don't be discouraged. Finish the book and you will see that when Sven poses a problem, he offers suggestions how to solve it. Some of his suggestions may seem a bit off-the-wall. Is a study of hydraulics likely to generate anything useful for managing projects? Maybe so or maybe not, but behind Sven's ideas is the big idea that our thinking should be more adventurous and our practice more thoughtful.

The inclination to disregard suggestions like hydraulics is one side of a two-sided risk. Disregarding 'wild' ideas is one risk. The other is to disregard those suggestions of Sven that may sound like common sense – don't select project team members on price, engage direct workers in managing the project, only count projects as successful that have delivered value to customers and stakeholders within cost, time and other conditions of satisfaction. That may be common sense, but how commonly are these things done?

Sven's vision for a science of projects
In his Postscript, Sven calls for the creation of a center to promote study of projects of all kinds, from construction to the performing arts. Like that unruly horse, his mind leaps forward into new fields with new challenges and opportunities.

You will enjoy this book; but be prepared for a wild ride.

Berkeley April 2016
Glenn Ballard

My life with projects

I love projects.

For better or worse, projects have been a way of life for me. Not least projects that went right to the limit and sometimes slightly further, because there so obviously lay new opportunities in challenging the traditional thinking and rules – with due respect to the objective.

My long life with projects has brought me many good experiences, but there have also occasionally been projects that just did not. Naughty projects I began to call them in Danish in lack of a better term for what in English is called wicked. This annoying about rushing project that again and again acts wildly and crazy and destroys my careful planning and my quiet night's sleep.

But over the years I came to understand that these projects were not wicked at all, they were not trying to make my life filled with problems as an act of bad will, and they just did so through their own behavior because I had not taken the time to tell them to behave otherwise. Just as when your dog does not follow your commands, or your horse or your child drives you crazy.

They may all be hyperactive but they are very seldom so and almost never wicked, they are just unruly.

But as creatures born to act civilized they are also tamable if you understand their nature and treat them accordingly, and that is what these essays are about.

My essays are therefore love letters where I try to explain the project's unruly nature and give some understanding of how to treat and train it to develop the kind and friendly behavior, we all dream about.

Throughout a long life, I more and more have come to see projects like living beings around me, a bit similar to my animals and children and indeed, my whole family.

For the same reason it hurts me when I see how projects everywhere are derailed because they are misunderstood. Or rather: Because the management has not understood the project's true nature and is therefore mismanaging these dear creatures.

A brief word about the English version of this book: Translating a text these days is simple, just let Google do it. But that is only the easy way to a readable text, but not to a text conveying the deeper meaning. This requires a diligent editing which Glenn has undertaken. During that process, we discussed whether we should move the text from its Danish setting into an English one but decided not to do so. Essays are personal thoughts, and in this case they are based on a lifelong personal experience in a Danish project producing industry and so it must be to keep its creditability.

It is then up to the readers, wherever he or she may come from to apply my thoughts on their own environment.

Introduction

We know it all very well, and if not we are reminded of it again and again – that we do not really have our projects under control.

Every day its own scandal is almost the motto, yesterday it was IT, today it may be new trains, equipment for army or air force and tomorrow certainly a large construction project. Indeed, especially construction is filled with projects that sputter away almost out of control, as Kevin Kelly calls it in his very entertaining book of the same name.[1]

There are probably few people who realize how big a part project production plays in our overall economy. But we're not talking about pennies.

Our annual Danish GDP in a country with app. 5 million inhabitants is roughly 2,000 billion DK Kroner, and it is hardly completely wrong to assume that about half of it is directly or indirectly attributable to projects. As far as I know the figure is not registered in our statistics, but it is a sober estimate based on comparable statements. After we have outsourced most of our mass production to the East, the vast majority of our domestic production has the nature of project production.

For this reason alone, you would think that large resources are channeled into researching and developing and improving our leadership of these projects, and our handling of the processes that damage this project production.

However, this is not at all the case.

We are talking about big numbers. Our construction sector alone stands for about 175 billion DKK annually. When I later tell about the waste we find in all project production, and in this respect claim that a productivity increase of ten percent is there for the taking in the kind of projects I personally know about, we are talking about a huge potential impact on our business, our project and our national economy.

An Australian study from 1992 showed that an improvement of the construction sector productivity of 10% would mean a GDP improvement of around 2.5%, which, if it fits the Danish economy, equals DKK 50 billion. This is the total price for the entire ongoing Danish super hospital program, every year, and with money for better research and training in addition.[2]

But now to the point.

These essays have been a long time coming. They are the result of a development project I started in NIRAS in the late 1980s – inspired by Toyota's production system and not least their just in time logistics – all much talked about at that time. But far more important were thoughtful colleagues, and the inspiration from architect Marius Kjeldsen (1924- 2004), who for many years had been and at the time still was The Ministry for Housing's thoughtful driving force behind the Danish development of the building process.

We tried in NIRAS with a group of dedicated players, with whom we worked in everyday building to see whether this Toyota logistics would work, and it did!

Far better than we had dreamed of, but despite a thorough planning process, we didn't understand why. We ended up with a productivity improvement of about twenty percent, and with a wide number of other improvements, but – as expressed in my later Japanese guru Shigeo Shingo's words – only with Know How and barely enough, maybe just with our own common sense and many years of experience with projects, but certainly in no way Know Why.

And so it went as it had to. In subsequent testing several of our enthusiastic employees tried to improve our very simple control system, and worse and worse it went, until the control was put into their IT system together with the schedule, and then it went completely wrong.

The whole thing was abandoned and went almost into oblivion.[3, 4]

In 1999, however, I found Lean Construction, where I suddenly found the same thoughts, and here they seemed to work. Certainly in California, but they worked. Sonja – my wife and sparring partner – and I traveled immediately to Berkeley, where we for the first time met The Three Musketeers, which I later baptized them. Lauri Koskela, Glenn Ballard and Greg Howell with their families became our personal friends, which we since have maintained close contact with.

Lauri Koskela is the thinker behind my entire understanding. Lauri thinks, sometimes so it creaks, and it is from his new visions that surprising angles appear. Who else could think to address the project's metaphysics with an inspiration from Aristotle?

It's his papers and articles and especially his TFV theory, which is the basis of my understanding of the unruly project, and especially for my discovery of the importance of a good theory.

Glenn Ballard, is the man from the engine room. He is at one and the same time a great inspiring and an alive and kicking front runner and an experienced practitioner. He is the man who started as a pipefitter's helper on a construction site and worked his way up to foreman and superintendent while studying engineering in the evenings. Later, he became PhD and professor at University of California, Berkeley.

Glenn is the man behind the Last Planner, the central system behind the new management of the construction process, and he is a researcher constantly looking in the corners and putting the relevant questions forward.

It is Glenn, who has led me to understand the importance of flow in the project.

Finally, there was Greg Howell. Greg is everyone's uncle. Engineer, Seabee officer in the US Navy and a Vietnam veteran, professor and thinker. Greg thinks broadly as I do not least about project management in the traditional sense. Like Lauri and Glenn he brings new trains of thought into the game, and usually spicing them with anecdotes from his own exciting life.

Greg made me look at complexity theory to explain the project's often irrational behavior.

Individually, they are three inspirational thinkers and as a group unique.

On this trip to Berkeley for the 7th annual conference of the International Group for Lean Construction – IGLC-7 – I found a new, different and deeper understanding of the project, especially in Lauri Koskela, who presented his ideas of a Transformation-Flow-Value (TFV) theory for the project. It was also Koskela who told me about the Japanese production engineer Shigeo Shingo and his thoughts, that are largely the basis for the Toyota Production System.

Also the control method Last Planner, which Glenn Ballard had developed with Greg Howell, and to confusion was like our Building Logistic System, but something they just made work. Our two systems resembled each other but somewhere there had to be a difference.

It took me some years to find it, and not least to understand it. It was the understanding of the concept of Flow, that clearly lay behind their thinking and which Lauri Koskela had put into place through his TFV theory. It had not entered our thinking. They had thus achieved a major step forward compared to us, because they, said with Shingo's words, had not only created a Know How, but more important a Know Why. Not full and complete, but they were well on the way.

My new friends had also found a good name for Glenn's control method: 'The Last Planner System of Production Control' or Last Planner colloquially, and they spoke English and could therefore be understood in most parts of the world, so it was Glenn who deservedly won the match.

And Last Planner has been successful. Today, the method often becomes synonymous with Lean Construction, which is too simplified. And worse: Last Planner has, because of its success, been embraced by well-intentioned 'improvements', as we saw with Building Logistics, but such well-intentioned improvements blur the understanding of Last Planner's quite simple message:

Create reliability!
Make sure that things can happen, when they should
happen, and that they then will happen.

The man on the spot – the last planner – is usually best placed to assess this precondition in the project's hectic everyday life.

Check subsequently that things really happened as planned, find the root-cause of every failure and remove it immediately.

The Message That Was Lost

Somewhere along the way in this long process, I felt an urge to get away from my consultant's thinking, methods and systems. The thinking is groundbreaking in its simplicity, and there is not in any way something wrong with the method itself. However, it seems to get closer and closer to being a prescription, which in each project is garnished with superfluous knick-knacks.

So with Shingo's words, it is shifting from know why solely to be know how.

In my seven essays, I now return to the problem: The unruly project where I try to build a better understanding of what it is we are dealing with when we launch a project.

So with Shigeo Shingo's wise words, once again create know why.

And what will I tell you?

In my **FIRST ESSAY** on the chaotic project I ask the fundamental question whether we know what we are talking about when we talk about the project. Lauri Koskela points to the strange fact that we have a huge literature on project management, but amazingly little about project production. Could it be here, that we must seek the explanation? Koskela himself offers his Transformation-Flow-Value theory, which is fundamental to my new understanding of the project. Here I allow myself, however, to adjust the theory slightly to align it with the project in practice, and to dig a bit deeper and find that there are quite a few aspects of the project's nature we have not investigated, and which are therefore not part of our thinking, understanding, organizing and managing.

At the same time I distance myself from the classic understanding of the project as a Newtonian clockwork, and chaos begins to threaten. Plans stop being followed, and we must think anew.

With this discovery, I look in the **SECOND ESSAY** critically on today's typical understanding of the project and thus the management of the complex, dynamic project as an ordered system, and what this management is missing, despite sophisticated control systems.

And again, I establish a new and broader understanding of the project's nature, and a far more inspiring basis for its management. At the same time, I examine the dynamic project and its fluctuations between the orderly and chaotic, as I explain with an analogy to the two states we engineers know from hydraulic science's treatment of water flow: laminar and turbulent. For no matter how reluctantly we as project managers want to admit it, when the project work best it balances on the ordered edge between these two states, but who remembers Reynolds number?

In my **THIRD ESSAY** on the fluid project I allow myself to introduce an often nasty notion for many practitioners, namely theory. I look a little into some of the many theories we have available in our new approach to project management. Especially engineering sciences such as hydraulics, complexity and management turn up here. Whether we like it or not, there are indications that we must learn to balance on the edge of chaos, if we are to manage the project properly. And here the social sciences also come into the field, for ultimately the project is all about people.

In the **FOURTH ESSAY** on the complex project, many may think that now I'm running completely off track. But don't worry, I am standing on safe ground when I introduce complexity and chaos in my understanding.

In the **FIFTH ESSAY** on the methodological project I move

from theories back to the unruly project and its everyday life, and I outline a new approach to the management of the project, as I today would approach it if it was a major building project. Not a guide, let alone a manual, but only considerations aimed at utilizing the theories in practice.

This leads me to the **SIXTH ESSAY** about the independent project to reflect on leadership in chaos, where I involve thoughts known from management in completely different industries, such as the art of war, where as you know, something unexpected always happens.

When I bring these considerations into play, it is because the unruly project by nature risks running wild in combination with the fact that we cannot just keep it steady with strict management. If we push too hard, our Reynold's number may exceed the critical value and our flow become chaotic. The optimal situation exists always on the edge of chaos, as I had noted previously in the third essay.

The **SEVENTH ESSAY** on the living project is hardly a surprising consequence of my new approach to project management: Leading the unruly project is not about order, discipline, contracts, plans or systems, but about creating reliability and cooperation, and thus retrieving the enormous productivity gain, hiding in the project itself for the benefit of all its participants.

And here I naturally arrive at some of the many institutional barriers we today have raised about the project and its implementation.

By this the story is told, but just because I explain it here, it hardly happens by itself, although there is an emerging trend around us. But there is much we do not know or where we do not connect our knowledge. On my way, I have roamed through much science beyond the project's usual universe – hydraulics, control theory, chaos theory, production theory, management theory and the art of war, to name a few, and I suggest therefore in my postscript how all this knowledge could be brought in play into our handling of the project in everyday life.

The seven essays are mainly based on construction projects. Not only because it is those I have met most in my more than 50 years as consulting engineer, but also because building and construction is a key element in the development of our society. But just the breadth of my own experience means that I can also see a general pattern in projects where I have personally seen the methods work, which is in construction projects, in the development of IT systems and in shipbuilding.

Essays are, as the word says, one's own speculations. They are told here in the light of a life's work on projects, and therefore may be a little chattering, but anecdotes are an essential part of our heritage and our retention of knowledge and experience.

In other words, this is not science, and my essays have therefore deliberately not been subject to a formal scientific review. They are and will remain my own thoughts.

For readers not familiar with Lean Construction I have established a homepage for the book, where I will place the papers to which I refer along with an introduction to my own understanding of Lean Construction and hopefully the readers' supplementary comments to my thinking as we are looking at a dynamic development.

www.theunrulyproject.com

1) Kelly, Kevin (1994): Out of Control – The New Biology of Machines, Social Systems, and the Economic World. Addison-Wesley Publishing Company

2) Stoeckel, Andrew and Quirke, Derek (1992): Services: Setting The Agenda For Reform, Prepared by the Centre for International Economics for the Service Industries Research Program under the Department of Industry, Technology and Commerce, Australian Government

3) This whole story is fully recorded in Bertelsen, S (1993, 1994): Byggelogistik - material management in the construction process Vol. I and II, the Ministry of Housing (In Danish)

4) Bertelsen, S and Nielsen, J (1997): Just-In-Time Logistics in the Supply of Building Materials. 1st International Conference on Construction Industry Development, Singapore.

The chaotic project

*Where I ask my key question,
whether we know at all
what we are talking about*

ONE OF MY CLIENTS, an experienced builder, spontaneously said some years ago, in the midst of the quality assurance reform: *Never have we had so much quality assurance in construction and such poor quality.*

Today we can similarly say that never have we had so much project management, while so many projects are running out of control. Everywhere we see it, and it happens far more often, but the stories are not told. And I am not talking just about construction, the highly visible projects, but also about projects often living in hiding. IT systems, for example, where you really can see runaway projects never even completed, but which must be abandoned without any other results than a consumption of unimagined millions. It rarely turns out as bad as that however with buildings.

This miserable situation is often explained by the projects

being more complex. Maybe it is true, but projects have always been complex and although they may have grown more so, we also have better tools to help us in their implementation. And there is much more literature on project management in the form of research, textbooks, guidelines, management-courses, certifications, management consultants and... Indeed, a whole new and thriving industry has risen around the projects' large and somber cemetery, where nobody comes to lay flowers, but everybody rush away and try to forget.

It may be because our projects must be completed faster, but The Empire State Building i New York, in 1931 the then world's tallest tower, was built in 13 months, and the whole project from the first sketches to the inauguration took 21 months. With materials from far and wide, marble from Italy, steel from Pennsylvania and a construction site manned by unskilled immigrant workers from Europe and Mohawk Indians from upstate New York, the building grew at a rate of one floor per day. No SMS or e-mails; telephone calls from state to state were difficult, and drawings had to be painstakingly made by hand, printed in a stinking ammonia process and sent by courier by train and ship over long distances.

But the building stands there, so we *can* manage projects.

Perhaps it is rather the other way around: The tools make us cocky, so we start increasingly more complex and dynamic projects, while we develop corresponding complex control systems, which do not really control, but often just add further complexity. So we create for ourselves a vicious circle. If it is true that projects are becoming increasingly more complex, then why do we not deal with complexity itself? Maybe not by eliminating it, but rather by systematically reducing it and, not least, accepting and understanding it, and thus making it manageable? For complexity we find everywhere, and today the understanding of complex systems has become a science that can really teach us something; a science formally called

Complexity theory and colloquially often *Chaos theory*.

> *I am not joking. I mean it in earnest; we must stop and think it all over again, and begin to understand the project as the complex system it is and always has been, and learn from the complexity theory that actually deals with chaos, but in an orderly and logical way.*

It is a great challenge because it breaks with nearly 500 years' thinking, and with our classic, rational and scientific worldview.

We are Stuck in the Renaissance World Order

Many of the problems we struggle with in our daily battle to make the project clean up its act, are simply due to the fact that our mental model is wrong. We are all – in our part of the world – brought up with the logical and rational world order of the Renaissance, which says that everything can be explained and understood, if only we think systematically.

My history teacher in school was very interested in the Renaissance, and his inspiring descriptions are still deep in my memory. How wealth generated through trade, with the art of calculation and navigation in the background, was assembled in the Mediterranean countries, and provided incentive to the great voyages of discovery that brought additional wealth. And how an increasingly confident science challenged the Church's worldview and brought the understanding of the universe into order, so that it was no longer the earth that was the center of the universe but the sun, and the planetary orbits were explained through Newton's laws, developed from Tyco Brahe's observations and Johannes Kepler's ellipses, and last but not least, how the art of printing, the revolutionary new information and communication technology of that time, made long-range cooperation and exchange of ideas far more effective than ever seen before.

The world became like a clockwork. If we understand the details, we also understand the whole.

I avoid art, as I cannot claim a deep understanding, and yet the perspective in Leonardo da Vinci's Last Supper was probably also a rational analysis of our world. And then sometime later Johan Sebastian Bach, whose Brandenburg concertos to a rational engineer may sound just as pleasant as the sound of a well-oiled machine, orderly, rhythmic and beautiful, if the clockwork is your ideal, but infinitely far from the East's soft flow of tones.

Perhaps there is here an explanation for Toyota's success in fully understanding flow? The German thinking about flow of production speaks of 'takt', which is also embedded in Ford's assembly line – and on the surface also in Toyota's production system. But my feeling is that there is a more flexible approach to flow at Toyota, and their approach to errors is decisively different. Toyota uses mistakes to learn from, while in the West they are often considered sins of the production that have to be hidden and quickly forgotten. Just think of the wealth of experience that may be buried in the pile of projects of all sizes that fail each year. Would it not be reasonable if we dedicated just a thousandth of the project cost to post-mortem studies through a systematic evaluation of what we learned, and especially what we learned from the projects that went wrong. Not criticism, but rather a humble learning process recognizing that we were on the wrong track.

Again a meeting between two very different mind sets on the basis of different world images. It's probably the same thing we see in the two different cultures' emphasis on efficient operations that we see in the West, and reliable flow that seems to be the criterion in the East?

The French philosopher Pierre-Simon, marquis de Laplace (1749-1829) formulated the foundation of our Western belief in project management in saying:

"The project's state today is of course a result of what it was yesterday. If we think about a project management that at any given time were able to oversee all participants and understand their relations in and outside the project, this

*project management would be able to show the project's
and all its participants' situation and activity and the total
operation at any time earlier, today and tomorrow"* [1]

Hello! Here is the description of the perfect project management stated more than 200 years ago.

I have here allowed myself not only to translate Laplace's statement - his demon as it is called - but also to modernize the language a bit with words from the project's life. But the meaning is clear: the World – and thus the project – is like clockwork. When all the pinion parts are in place and oiled, it runs along just so and the project management can, so to speak, turn the button forward and see where it will be in a month or two – right up to the planned conclusion.

It is this mechanistic view, that everything can be predicted, which has led us to today's belief in planning and management – and not least to the belief that plans can be followed, which in my understanding is the fundamental error in today's project management.

*Plans never hold. Not because of bad planning but
because plans cannot hold in the real world!*

The Devil Is in the Detail

De Laplace had, in all his logic ignored a single little detail, like most of the thinkers of his time.

Since Newton's laws were simple and beautiful and easily could explain the moon's movement around the earth and the earth's movement around the sun, it was assumed that they should of course also be able to explain the entire universe's movements with beautiful equations. In other words, the world could be explained as a mathematical system, where you just had to solve the equations. Admittedly, this was not so easy when both the sun, the earth and the moon were included in the system, but then one could at least step by step calculate the celestial conditions as de Laplace said. Laborious but logical. It was known that the real truth was to be found in the non-linear equations found going from two to three bodies, i.e. the Sun, Earth and Moon together, but in practice this kind of equations were almost impossible to solve with the methods and tools then available, so scientists simplified the problem and assumed that the small inaccuracies which thus arose, hardly mattered, they were just looking for the bigger picture.

And by and large, it seemed to work, and the world seemed predictable, although the plans still had difficulty being followed.

When the Butterfly Brought Chaos Back

The American meteorologist Edward Lorenz (1917-2008) was proud of his new acquisition at MIT in 1961: A digital computer, something new and quite exciting.

It was in the years when these 'electronic brains' escaped from their hitherto captivity in special laboratories and basements, and with transistors instead of tubes and valves started to develop rapidly toward today's PCs, tablets and mobile phones. But at that time it was magic. Now ordinary academics could write programs for these machines in an understandable mathematical language, because visionary engineers already had written other programs that could translate this mathematical text into the machine's demand for a large number of zeros and ones: 10011101100011 etc. Now it was

```
Xnew:= X + 10*(Y - X)
Ynew:= Y + X*(25 - Z)
Znew:= Z + X*Y - 8/3*Z
```

Actually the equations express – greatly simplified – the balance in the atmosphere described by X, Y and Z, and they show how the state will be a time-step forward. Simple on the surface, but the equations cannot be immediately solved, because they so to speak are biting each other's tails and therefore form a dynamic system. Going one step further to Xnew,

Y will change and Z as well, and X is changing as a consequence.

For non-mathematicians one could perhaps explain this system of reciprocally dependent elements by expressing the equations through a story about three children, Sanne, Tom and Sean playing together. Basically, their situation is X, Y and Z, but let's suppose the equations have Sanne's situation changed with ten times the difference between Tom's and Sean's current situation. It may be that Tom and Sean are fighting over the swing, which affects Sanne's chance to swing. Tom's new situation arises from Sanne's desire to swing, but also influenced by Sean's interference, and Sean himself is affected by the combined resistance of Sanne and Tom, but reduced by his own resistance.

Maybe not simply formulated, but when are three children's turbulent play simple to predict?

The phenomenon had occupied mathematicians for centuries, but they had for many years just pencil and paper to use – and got nowhere. Edward Lorentz, on the other hand – who saw the equations as a simple model of the weather system – now had an electronic computer available and with its help he expected to calculate step by step his way through the system's development. We recognize de Laplace's thinking – predicting tomorrow's weather from today's. Of course, Lorenz knew that he did not have all the elements in this very simple model with the three equations, but he expected that the small inaccuracies would remain small because they were small and innocent at the outset, so in the greater picture they would be inconsequential.

He loaded the formulas into the machine and checked that it was running properly and sat down and waited. Now computers at that time were much slower than even a mobile from before we got smartphones, so it took time. And equally important for the chaos theory's emergence was that the output from Lorenz' computer came on a so-called teletype machine, an electric typewriter, writing the results on a roll of paper.

All this took time, and while the machine faithfully gnawed on the three equations, Lorenz, walked down the hall to get a cup of coffee. When he came back, the printer paper roll was however used up and a new one had to be put in and the calculations started again. For good measure, he started the machine not from the results, it had reached, but went a number of steps backwards to create a solid overlap. And then he started the computer again.

Great was his surprise, however, when the overlap did not match, which quickly became apparent. The new forecast showed in no way the same results as the first one had produced.

Mysterious! He had re-started at a calculated point and with the same formulas and rules, and yet the two simulations did not proceed together. It was not nature or humans he looked at here, but only math and formulas, so where was the mistake?

In the computer, of course, for it had deceived him when he restarted after changing the paper. It had saved a few more decimal places in in the far end of the results than it had shown in print-out, so Lorenz did not start at exactly the place he had arrived at, but with a very small deviation.

What Lorenz had found was not really new, as mathematicians around him with some right claimed. His model, simple as it was, simulated what they called a non-linear system, which belonged to a class of problems they well knew existed, but that they in general had let lie, because they were so difficult to tackle. One perceived them as curiosity cases and approximated them in practice to linear systems that gave equations that could be worked with.

The rational worldview broke through again, although there were some skeptical voices.

Lorenz wrote an article on his discovery titled: Can a butterfly flapping its wings in Brazil set off a tornado in Texas?[2]

However, he could not get the article published, neither

among meteorologists nor mathematicians. For mathematicians it was something old; for the weather forecasters in contrast it was far too new.

He managed to do it only ten years later, but there had actually already been a few mathematicians who had smelled the problem much earlier:

It may happen that small differences in the starting point causes large differences in the end result. Small mistakes do not remain small, and prediction becomes impossible. [3]

So clearly and precisely the French mathematician Henri Poincare (1854-1912) buried the foundation of de Laplace's understanding of planning already in 1882. And that was what Edward Lorenz rediscovered when he went for a cup of coffee that night in 1961 at MIT in Boston.

Chaos was rediscovered!

As Lorenz put it himself:

Two states, infinitely close to each other, can develop to two completely different modes – and in any realistic system mistakes seem to be inevitable – a reasonable prediction for a future state may well be impossible. ... Precise [weather] predictions do not seem to be possible.

The Project

All these new-fangled ideas are surely only theory, many thought. In real life situations such as in construction with strong craftsmen with hard hats and big yellow machines, one could certainly keep things on track, although there were a few butterflies here and there. So the industry continued as before, and that is virtually the case still today, fifty years later.

If we look at today's project production, we therefore still find Newton's and de Laplace's thoughts. The project can be planned with a higher or lower level of detail, so there is a plan to follow, which everyone tries to do. The project is regarded as a clock work, in which all the parts must be understood and developed according to the plan, and then it will work as it should do. In construction the project is divided into subcontracts, which are delegated to subcontractors that are expected to work as the plan says they should. Architects, engineers, craftsmen and suppliers are all selected by lowest price, with the logical expectation that the lowest price of each part result in the lowest total costs.

If the budget otherwise is kept with all these lowest prices the project moves forward as the plan predicts, and what happens then? Exactly the same as in Lorenz 'simulation: The project does not follow the plan.

Oops! That was not good, and so the plan is updated and we are back on schedule. The involved participants are talked to seriously and all say "yes sir" and try to align. But the weather is all its own, and there is also always so much more unforeseen and often something quite unthinkable. And a week or two later the plan does not fit again. Then the whole show repeats itself, now spoken in slightly larger letters. All is aligned again, the plan is updated and it all continues, while the money flows out of the till and time progresses.

Along the way new and more advanced so-called control systems will be put into operation, the project will be reorganized and replacement in key positions will take place, the mountain of paper will grow and meetings will be held all over the place, but nothing seems to help. As the money is pouring out, the ambitions of the final result are reduced, while prices are pressed once again. And time passes.

Everybody discusses, cries and argues, but it seldom helps. The unruly project has its own life and the plans may say what they will, it does not bother the project one bit.

The Systems

I have in my long life as a project manager met many of these magic tools that should save the world and help my project through, and which I then believed in – from the 1960s'

PERT, which today is almost forgotten, and the simultaneous CPM still alive and well, partly as MS Project with variants in different clothing. In scheduling, we have also used Gantt maps – among general users called bar charts – and later the new tool in the box: The good old Line of Balance, also called Location Based Management or simply Cyclogram, an old and proven method for making train schedules.

Bar Charts and CPM are based on the simple assumption that the project consists of a number of tasks – operations I call them later – to be performed in a specific order to reach the target. Just as a recipe in a cookbook. PERT does apparently the same, but with a small and usually neglected difference, for PERT does not look at the operations, but at their prerequisites, and thus on the flow. But this was overlooked by most, and today the method is probably only history.

CPM made much more sense because here you looked at the operations, i.e. the contracts, where so to speak something happened and the money was spent, and then there were all the IT systems that supported the method. But they overlooked the fact that there are other prerequisites for an operation to be executed than merely that previous work is completed – for example space and manpower. Here Line of Balance came in and interconnected these three things in the so-called Cyclogram, which today is considered by many to be *state of the art* in the field of project management.

But digging down behind the thinking, it's still the understanding of the project as an ordered series of operations that reigns. Newton's clockwork – now with dolls that move by hour strikes as Bavarian tower clocks.

In construction the project progresses normally forward through this chaos where no one seems to have control, and the project usually reaches an end with a usable result. Rarely absolutely the best, but, however, usable, and even if the budget was exceeded and the result was delayed, it was at least completed and put into use.

Elsewhere, it often happens that the project is terminated without any useful result. It ends just because now we don't bother boxing with it anymore, and moreover we have found another solution. Just ask about in the world of IT.

Nobody seems to take Lorenz seriously.

If we turn to project management's thinking and actions, we find that notwithstanding the term, project management is in no way managing the project, it manages plans and contracts only in the belief that if all parts of the clockwork are doing as they should, the tower clock works too.

Which it rarely does.

Must it Really be So?

This is where I stop and ask whether it really needs to be like this? Must projects end up as nightmares for all involved, or at least for some of the participants. Must the project be a war with numerous victims and suffering, from which only a few return pleased and proud – if any at all.

The project today is like a war. Not against an enemy, but against the unexpected, the unforeseen and the unlikely.

All this we may to some extent change, but only to a certain extent. Solving the problem requires that we stop completely and rethink our whole understanding.

There is something fundamentally wrong in our understanding of the project we struggle and harass with, but it's like a living creature, that just will not clean up its act and behave like a Newtonian clockwork, maybe with a few small and innocent mistakes.

So rather than correct and improve – not knowing if I improve through my corrections – I'd rather wash the blackboard clean and start all over again, from the hypothesis that our prevailing understanding of the project is inadequate.

The project is not at all like what de Laplace and Newton and all the other thinkers believed, nor like we learned in school. Complex systems are not orderly and predictable as the machine, but rather chaotic and unpredictable, as Niels Bohr surprised Albert Einstein by claiming in his quantum theory, and Lorenz discovered with his weather forecasts.

So, let us – rather than fight with this unruly creature – stop and try to understand its nature, before we talk about how we can possibly tame it.

To put it briefly, before I go on in my thoughts on the project, its nature and its management, I believe that our rational understanding of the project and of the whole world that surrounds it is wrong. Not just a little wrong, but fundamentally wrong, and that this wrong understanding is like mold, which spreads both up and down and sideways throughout our handling of the project.

All while we suffocate from the number of projects being derailed.

Plans never hold. Not because of bad planning, but because plans can not hold in the real world.

Go and See

If you put on your gumboots and hard hat and go out to look with your own eyes at the wastage taking place in every building project every day, all of which apparently looks like something quite natural, you will find that only a third of the worker's time is used to actually build. Another third goes to prepare the work, while the last third is just waiting.

This very fact alone, that it is only one-third of the time that we build, should make us stop, think again, understand the unruly project and find new ways to rein in its whole dynamics.

---

1) de Laplace, Pierre Simon: A Philosophical Essay on Probabilities (1814)

2) Edward, Lorenz. Predictability: Does the Flap of a Butterfly's Wings in Brazil Set Off a Tornado in Texas?. American Association for the Advancement of Science, Washington, DC (1972).

3) Poincare, Henri (1903): Science and Method

The dynamic project

Where I speak about the forces that are in play in the project

IN 1991 THE YOUNG FINNISH ENGINEER Lauri Koskela asked almost the same questions as we in NIRAS had asked ourselves some years earlier, whether the modern Japanese production system could be used in construction.

Lauri Koskela was – and is – however a researcher, at that time studying at Stanford University in California, so the result of his considerations was therefore a short analysis written in 1992 and later his groundbreaking PhD thesis from 2000. [1,2] In this work he noted that while there existed much theory about mass production, the production of projects was remarkably little studied. In fact, there was not an overall underlying theory for what the project was, and how it consequently should be managed. Quite remarkable in view of the great amount of literature on project management there existed already at that time.

Koskela looked at how the understanding of the concept of production in the manufacturing industry had developed since the days of Adam Schmidt and Frederick Taylor, and he found that it roughly had happened in three phases. Originally production had been seen as a series of process steps – transformations – where materials changed shape and gradually grew in value towards the finished product. Later came the understanding that production should also be seen as a flow, where the product is moved through the production system in a continuous chain of transformation, inspection, transportation and waiting. All these four kinds of operations triggered costs, but only the transformations contributed to the product's value.

It was with the same understanding that Shingo had claimed that there were two kinds of production activity: The value-adding ones – and waste. Some waste is perhaps necessary, Shingo said, but waste it is anyway, and waste should be eliminated from the process if it should be made more effective. This was the foundation for Toyota's efficient production, where precisely flow and reduction of waste are the keys.

Finally, towards the end of the 20th century came a third discovery, namely that the production is also the creation of value. If it creates no value, the whole production itself is pure waste.

It was these logical arguments Koskela put together as a basis for a new understanding of the project, known today as the Transformation-Flow-Value theory – or the TFV theory.

This was a key discovery for the development of a new and better form of project management, known as *Lean Construction*.

Koskela chose for historical reasons to keep the word *Transformation* where Shingo's name Operation in my understanding is better from a production perspective. With the term operation, we can see the non-value adding activities such as inspection, waiting and transport, which are also elements of the process. And since the creation of value so obviously is the most important thing in the project, I have decided to rename the theory: *The Value-Flow-Operations theory* or the VFO theory.

Hitherto project management had only looked at the orderly sequence of tasks – transformations, activities, operations, trades or whatever they were called. Now suddenly both flow and value came into focus, and the whole ended up in a beautiful pattern.

The project's eternal triangle fell into place, and thereby a completely new basis for its management was born.

The Project's Eternal Triangle

The project unfolds itself at the intersection between creating the desired value, doing it to the right time, and doing it within the given economy. These three aspects are always to one degree or another in conflict, which naturally leads to the fact that they must be understood, followed and managed separately, but that it must be done as a coordinated whole.

You need to know which button you turn when you adjust – and know how this affects the other two parameters. This dilemma was not new, but now there was a logical approach to its treatment.

Value

With the VFO approach, it is immediately clear where the value belongs. Understanding and managing the concept of value is in my opinion probably by far the most difficult in the project's life, especially because we do not have an objective measure of value – in the same way that we do not have it for beauty. In construction, there is a whole science on this subject, namely architecture with its own universities and research. The classic understanding of architecture often takes an outset in the Roman architect and engineer Vitruvius' masterpiece de Architectura and look at *Utilitas, Venustas and Firmitas* – Utility, Beauty and Durability, as the perspectives architecture should be assessed under.[3] And I myself interpret the perspectives as it is the users who are thinking of usefulness, the world – all of us – that look at the beauty, and the owner who looks at the durability.

But that's not all. When we come to the project there also exists a time-element in value. In this view you may have one value perspective while it is carried out, a second when it is finished and the result is commissioned, and finally a third for posterity.

Especially in Europe we have a predilection to retain and refurbish our houses and maintain our cities. But when should we preserve and when should we pull down and build new again? A balance of value between modernism and museum, indeed.

Other projects may have different weightings between stakeholders and the three aspects of time, but as I see it there are always the nine perspectives in all the types of projects I can think of, from warfare, change of an organization, development of an IT system, shipbuilding and construction. The three groups of stakeholders will always be there, and the three time-perspectives as well, that's life.

But there is a trap hidden here. We are dealing with two very different kinds of value. When the real estate agent thinks about a house's value it is its trading price, while the owner may well have completely different thoughts; namely its user value. It can be good neighbors, the children's playmates or a particularly good apple tree in the backyard, something which is often difficult to formulate and impossible to explain, let alone price and put into the sales presentation.

More theoretically one can speak of *Value in Use* and *Value in Transaction*, i.e. utility value and market value, and this is where we experience the problem in our daily lives.

In the Danish Ministry of Housing's last major initiative, Project House, the Ministry's young secretary Gert Vig in 1999 suggested provocatively the headline *Double Value for Half Price*, and challenged thus all branches of the construction industry to participate in a rethinking of the construction process. It was in years of optimism, and it was easy to establish the ten working-groups, undertaking the process of formulating a program for a 10-year development effort, especially driven through the subsidized housing projects, just as the social housing sector had been the engine of the industrialization after the second World War – and had then shown its ability to push productivity an enormous step upwards, enhance quality and make it economically possible for the unskilled laborer to move with his family into one of these modern homes, he himself had been a part of building. The Ministry for Housing had a solid history to build on – and had at the same time visions.

And then it started. For more than a year about 150 dedicated leaders from all branches of the industry bent over this challenge to rethink the building process.

Almost from the first day however, the process ran into the problem of the definition of value and not at least *Double Value*, which had not been defined beforehand. Where the intention from the start probably had been value in use, a significant real estate agent turned it into market value, and lot of creativity was lost.

Later several attempts to pick the subject up again were made, but I still fail to see value management in practice through the entire project life cycle.

Flow

In other of the Project House groups however, a lot happened, not least in the working group that worked with industrialized processes. The chairman here, Peter Henningsen, a Højgaard & Schultz vice president, was familiar with the experience from Building Logistics and with NIRAS' work in that field, and we found the Lean Construction method Last Planner. Together we set up a trial in a large housing project, Charlottehaven in Østerbro in Copenhagen, and we made a deal that NIRAS should help them on a *'no cure no pay'* basis, where we got no remuneration if the method did not work, but in turn could earn up to twice the normal fee if the method worked as well as we argued.

So we started, and to everyone's amazement – including our own – it worked from the first day. The threatening delay was negotiated, money was made by both gangs, subcontractors and the general contractor; and the client, an experienced international developer, was deeply impressed. Everyone was happy and it was with a smile that we were paid our double fee.

The Operations Had Been There all Along

The operations were the concept always known. They expressed the tasks in the project, which were procured from each participant – in construction the bricklayer, carpenter, plumber, painter… They each delivered their part to the party, and together these parts created the finished house, which then hopefully lived up to the client's expectation of value. It was also where the money was spent as each participant was paid for both work and for deliveries of materials and the use of equipment, and this transaction oriented thinking was the basis for the project management through systems like CPM.

What Was New Was the Flow

While Value was not seen as a real new element in project Management – architecture was after all a classical discipline in construction, and operations was what we had always controlled – the perception of the process as a flow gave surprising new insights. And it is here we found the decisive novelty in Lean Construction – just as it was found by Toyota many years before.

Here we deviated quite crucially from the classic management's focus on the individual tasks and instead started to look at their context. Not the momentum in the bricklayer's work, but what he could 'release' of finished work for the carpenter to work on. Reliability became an issue that got a whole new meaning. In the passing from one trade to the following it was no more a matter of 'almost'. Either it was ready for the next, or it was not. Almost ready meant the same thing as not ready, as in a fast ballgame such as basketball in which only the exact pass counts.

Now building was no longer a simple sum of the individual workman's operations, but a team-work where his ability to perform his own duties was obvious, but where his interaction with the other trades was the key issue. In everyday life arose concepts as 'done' and 'done-done' as a sign that the crew performing the task, volunteered 'done', and the receiving team volunteered 'done-done', when the result was received and approved as being in order, so that the next operation could start immediately, if the other preconditions also were in order.

This thinking naturally led to delegation, because in practice it is only the man on the spot, the next skilled craftsman himself that can ascertain *completely* finished, so that he immediately can begin. And this led naturally to collaboration and coordination at the lowest levels.

The social sciences came into focus.

The flow thinking naturally led to a new form of management, namely that a piece of work did not start because the plan said it should, but when the workers themselves were satisfied that everything was clear – that the previous work really was completely done-done.

This understanding moved the focus from the previous work to the total flow of prerequisites to start a task. After the flow understanding it was not enough that the bricklayer was finished. There should also be cleaned up and room to perform the work, the drawings should be present – the right drawings mind you – and updated to the latest version and the carpenter should be ready with the right skills and the necessary equipment and materials. And finally, all the external conditions – approvals, the weather and all that kind of prerequisites – should be in order. Altogether seven different groups of preconditions: Previous work, space, information, crew, equipment, materials and the external conditions should all be in place – fully in place and finished – before work could start. It was called that the task was sound.

It is obvious that nothing can be said about the duration of a task if one or more of the prerequisites are missing. However, the classical project planning and control is based on the principle that you prepare a solid plan based on assumptions about the sequence and duration of each task. And that you then follow this plan in the same way as the railroad trains follow their schedule.

Studies have shown, however, that this does not happen in reality. Each of the seven kinds of flow represents far more individual flows, in the construction project probably about fifty on average for each task, and this means that only a little bit should happen before the task becomes unsound. In common project performance, it is believed that even in the most controlled, lean project it is possible to obtain an average probability of 95% of each of the seven categories, which means that the probability of its total soundness – the plan's assumptions – rarely exceeds 70%.

In a normal, well organized and managed project the reliability probably is about 90% on each of the seven flows resulting in approximately 50% soundness on the task. And actually that uncertainty can be observed in the weekly work plan, having reliability around 50%. But this kind of observation is not common practice, and in any case it rarely happens that anyone is thinking all this uncertainty through to the end, and does something about it.

Everybody just believes that the project will follow the plan, the damned thing just has to do so, but that it just does not. On the contrary, again and again it finds surprising new and unexpected acts of wickedness. The truth is that plans are never followed. Not because they are bad, but because they can not be followed!

But the flow understanding gives much more than just a focus on the hand-overs in the process. It also opens up the door to a whole new universe to the understanding of the flow, which we find in physics, namely hydraulics.

Hydraulics

Hydraulics is the engineering science of the behavior of fluids, but many of its concepts can – with some caution – be used in other forms of flow and thus the work flow as well. One such concept is Fluid Mechanics.

Toyota early demonstrated the importance by moving their eyes from the operations to the flow. Should the flow in their production be effective with all the side-flow that obviously exists, the entire production system must be reliable. Earlier they had dealt with the inherent variability by buffers stocking spare parts for the assembly line to continue to run even if an error occurred, but now they removed all these small reserves and instead focused on reliability. Things just *had* to be in order – otherwise the assembly line stopped!

Oh my God, the assembly line must never stop, was the motto otherwise in car manufacturing, but at Toyota it might. For if there was an error, its root cause should be found and removed, so that this error never came back. Only in this way will we be error-free, clarified Shingo.

And so they did, and succeeded!

There is Always a Bottleneck Somewhere

With all this talk about flow, Glenn Ballard early in our relationship drew my attention to Elihaou Goldratt (1947-2011) and his *Theory of Constraints,* the bottleneck theory.

Goldratt's key word is *Throughput*, that is how much finished work comes through the production system to customers who *pay* for it. This is a purely profit consideration: How much money comes in as a result of what we do? A clear and precise flow thinking.

Here, however, he asked the question: Why have we not a higher throughput – and found the answer in the form of a bottleneck in the system.

In his first book on the subject, it turned out to be a robot to enhance the efficiency of some of the operations, and this it did. But the robot had been a great investment and should therefore be used optimally. This meant that at all times there should be tasks ready for the robot, and the variation in the flow caused inevitably a queue braking the flow – and thereby reducing the throughput.

The efficiency was increased locally, but the throughput and thereby the productivity – and consequently the earnings – fell.

Difficult to understand if one looks at productivity improvement as a sum of many small efficiency steps, but productivity gains are created by improving the flow and not by many small savings, a misunderstanding I repeatedly encounter in project producing companies.

Their financial system for example shows middle managers as expenditures and here they endeavor to save. But good middle management is a key condition for the reliable flow, because these managers should aim at ensuring sound work packages. But now they are congested and spend most of their time 'firefighting' and are therefore not preparing the coming tasks, which in turn cause new problems.

And all the time something is waiting for people and machines, the efficient capacity drops, productivity falls and the delay must eventually be overcome by expediting with high extra costs.

You cannot save to profit.

It is the bottleneck, which determines intensity of flow - and thus the throughput and productivity of the entire process, creating profits!

Hello! Is it really that simple?

Both Yes and No, but in principle it is. When we recognize that only a third of the working hours on an average construction site create value, there must be quite a lot of capacity that can be brought into play by increasing reliability and ensuring sound tasks. And this can be done almost for free, because we already pay for all the hours. With smooth and reliable flow of prerequisites we easily increase throughput with ten to twenty percent, it's only a matter of moving four to eight percentage points from the nearly seventy unproductive, then there we are. Construction time goes down, precision in production increases, we make fewer errors and work accidents decrease.

Corporate profit and the crafts' piece rates will also increase dramatically. Naturally, the craftsmen's piece rates will grow by the same ten to twenty percent, and this is where we see the first indicator of the methods working. For a skilled craftsman paid by piece rate always knows how it ends, long before anyone else.

The contractors are often more skeptical: The entire gain is going to the workers in that way, they say. But they do not understand the importance of flow in their own business, because they have never calculated on the simple equation: What if we increase the flow – i.e. the throughput – by ten percent?

I have presented the calculation several times and have done so in a popular form in my book Semiramis, but also more scientifically in an IGLC paper. The result surprises most people. It can't be so, they say. But it is. I've seen it again and again, it just happens, if one gets into the thinking behind and is willing to unlearn old misconceptions and dogmas. A doubling of an already positive bottom line is almost inevitable, if you focus on the flow in the right way. [4]

And it happens almost spontaneously!

It applies not only to the construction process.

Some years ago there was a pause in the meetings on a Norwegian shipyard, where I helped introducing Lean Shipbuilding. We walked through the dock hall, and I asked that everybody mentally registered every time they saw a worker, what he was doing right at that time.

Did he work – create value, did he prepare for working or

was he waiting? No one should write anything down, just observe and not exchange observations.

After half an hour we stopped and compared our observations. All had found nearly the same: In one in four of the observations, the worker was in the process of creating value. In every third he was preparing to work. The rest were waiting.

Then we walked back the same way, but now we stopped along the way and talked about what slowed the flow. Now pictures were taken and ideas exchanged. The result was a regulation of the traffic up and down the two narrow ladders. Up in starboard, down in port, a double wide gangway with room for passing traffic, decks cleared of cables, and a little later a construction site elevator from the pier up along the eight-floor high deckhouse.

And then a whole bag with additional ideas: Service areas near work stations with restrooms, a coffee and drink vending machine, information kiosk with printer; tool and material containers on the deck on board the vessel and not on shore.

Productivity and timeliness increased dramatically, also because the Last Planner thinking was introduced and the yard was soon setting a completely new standard in a highly competitive market.

The Bottleneck – the Devil in the Game

In any flow system there is one and only one bottleneck which determines the flow intensity and thus the throughput. When the throughput has to be increased, it is this bottleneck we must find and ease. But where does it hide in the seven flows that create the sound work package? All seven are usually extremely complex, just going a few steps upstream. While they usually are quite clear on arrival for the job, they soon mingle upstream. Delivery of materials is waiting for information, often drawings, which is waiting for other information; external conditions such as permits for materials to be bought and delivered cannot be obtained without drawings and specifications – and while this takes place all are waiting.

Seen from each task it is simple: There are seven kinds of prerequisites that must be in place for the task to be sound and there will always be one that is critical, i.e. the one the other six are waiting for until the activity becomes sound and work can begin. If we want to increase the intensity, it's here we must improve, it is the critical flow, so to speak, that controls the velocity in our production.

This is exactly like the stove. If we want it to heat more, we must either feed it more air or more wood, but it doesn't matter which of the two we do. Missing air is not helped by more firewood, and vice versa if it is wood that is missing. And neither helps if it is a sufficient temperature that is missing. In other words, three flows of prerequisites: fuel, air and heat, and one of them is the critical one, the one regulating the process. The heat is usually the result of the 'previous work', that is, if there was fuel, air and heat enough present a moment ago, so the three flows are mingled together in a complex pattern.

If we look once again at the project, we find the same, but far more complex. The seven flows weave in and out between each other and in and out of the project and every prerequisite, regardless of type, is the result of an infinite network of associations of mingled flows. We can safely describe this as an extremely complex system, and it is usually quite dynamic as well.

If there is not something unexpected happening in our project, it happens certainly in one or more of the numerous other projects with which we, in one way or another, are sharing flows.

No matter how much we try to sort this system, we may confidently call it chaotic!

Chaos

When for just over 30 years ago, we bought the family's first home computer – a VIC 20 with 16 kb of memory and a Basic interpreter – our son and I became, like Lorenz twenty years earlier, excited. The non-linear systems entered our world.

Now with the computer we could deal with these systems with a relatively high efficiency, and as Lorenz had discovered it was a whole new world opening up. In a series of very inspiring television broadcasts the then young natural science wizard Tor Nørretranders explained this world, and Rasmus and I threw ourselves straight into testing all this. Suddenly Lorenz' butterfly flapped around on the screen and Edward Mandelbrot's wonderful flower was unfolding on the computer, which was quickly replaced by the more powerful Commodore 64 and later an Amiga. There came also a PC into the family, and infinity, complexity and chaos were studied early and late.

Life was never the same again.

All of this also changed my understanding of the everyday process around us as well. I said with conviction in our management team that NIRAS was a self-organized chaos, and that projects correctly should be perceived as complex, non-linear, dynamic systems, and that they therefore had to be dealt with as potentially chaotic.

Potentially! For it was not always they developed as such. Some projects were tame and predictable, at least a good part of the time, but then the devil could suddenly grab them and let them gallop wildly over the plains, apparently completely out of control. Although I actually had all the elements, it took me a long time to put the pieces into place and explain this strange behavior.

But if we start from the VFO theory and focus on the project as a flow, we know from hydraulics that flow has two in principle different states: laminar and turbulent. In nature, we find the laminar flow in the lazy river, which steadily moves along and is quite predictable. The turbulent state on the other hand is found if we turn to its tributary, the feisty wild creek that leaps between rocks and boulders. It's the same substance, H_2O, we are talking about, but it behaves quite differently, and there is no state in between, it is an either or. Turbulence may occur locally, and either it spreads or it dies out, which can be a delicate balance. But the phenomenon is there, and it's the same thing we experience in the project flow.

Turbulence and chaos are therefore key elements of my understanding of the nature of the project – and thus an essential prerequisite for my approach to the management of complex and therefore potentially chaotic systems.

So let me for the record clarify what I mean with the term chaotic, which I consider to be a subjective concept:

The system is chaotic, when it is unforeseeable in the time perspective you want, with the accuracy you need.

1) Koskela, Lauri (1992) Application of the new Production Philosophy to Construction, CIFE technical report #72, Stanford University, September 1992

2) Koskela, Lauri (2000): An exploration towards a production theory and its application to Construction, VVT Technical Research Centre of Finland

3) Markus Vitruvius Pollio: De Architectura de Architectura - ten books on architecture, probably written around the year 15 BC

4) The model is presented in the article: Bertelsen, S and Bonke, S (2011): Transformation-Flow-Value as the Strategic Tool in Project Production, IGLC 19 Lima, Peru

The fluid project

Where I dive into hydraulics – and find that flow is not only relevant to Liquids but to projects as well

HOPEFULLY IT IS CLEAR TO THE READER still on board, that it is my hypothesis that our usual understanding of the project is fundamentally wrong, and that this largely explains its unruly behavior.

The fault lies in our own rational thinking, in our blind belief that complex systems can be called to order and follow a plan. The same mistake as de Laplace committed, as Poincaré realized and pointed out, but which took Lorenz approximately a dozen years to convey, despite the fact that modern computers made it easy for everyone to observe.

But what do we put in its place?

Project Physics

A dozen years ago Glenn Ballard made me aware of the book Factory Physics, written by the professors Hopp and Spearmann.[1] In a monster piece of formulas they walk through the factory production as a flow system and collect a lot of math that may help to design and optimize. Exciting, but all directed towards the ordered factory environment, which may not be as ordered in fact as the professors assumed.

But a very thoughtful book, indeed.

It caused me to consider the difference between mass and project production, as many others had done before me, and probably will do later. Basically, most us see the difference or maybe not. Even when I tried to explain it, I ended up understanding that it is not a question of either or, but rather both and, and as such all about a personally chosen point on a scale between the two forms in their completely pure forms, and thus to some extent a subjective view.

It became especially clear to me when I brought many every day, small projects into my thinking, where Sonja's warm pork rib roast with crackling skin for our Saturday lunch for her is a production, but for me would be a project.

This led me to the idea that perhaps it is rather our mental model – how we view the task – which determines what we call it, and therefore choose our approach and the methods we are going to use.

This realization made me suggest to Lauri Koskela that we should try to assemble a corresponding theoretical understanding of the project under the parallel name Construction Physics.

The idea was in its simplicity, that if factory production could be put into formulas, the project production could be done the same way. However, that could not be done at all, at least not by us. It was too unruly, but yet we learned a lot. We found among other things, that it was the complexity of the project that was in the way, as we have had suspected right from the start. Complex systems may not meaningfully be described in a set of linear equations, because they almost by definition are non-linear as Lorenz weather model.

And to this came the project's inherent dynamics.
But we took some perhaps decisive steps before we stopped,

and these first steps are the starting point for my further work on understanding the project, in theory as well as in practice.

Therefore, I intend to use the bulk of this essay to rethink my understanding of the many elements in Project Physics, which I now have begun calling the subject, and thereby present my own theoretical understanding of the project, as it is suggested through my first two essays. Theory Based Project Management it could also be called, as an outline of a science of the project's nature and its management. Maybe it's an ambitious aim, but I am trying anyway. So let me as a start quietly try to collect some of the elements, which in the light of today's knowledge should be part of such a new science.

An overview

First, a brief overview of the new understanding of the project, which I then will elaborate upon with existing knowledge from other disciplines, with speculations on new interpretations and with a few new ideas thrown in for good measure.

THE PROJECTS OBJECTIVE IS TO CREATE VALUE: The purpose of every single project is to create value for somebody, and therefore the delivery of this value should be the key issue throughout the project's life cycle.

THE VALUE IS CREATED THROUGH A FLOW and the velocity of this flow determines the duration of the project, the time, which is the second requirement in the project's eternal triangle we constantly contend with. The flow draws in a number of operations that trigger the cost. Here we find the third requirement to the project: that we comply with the budget. The operations may contribute to the value, or they may be a waste – but they all use money and take time.

THE PROJECT IS A COMPLEX SYSTEM: The project exists in an infinite network of players – agents they are called in complexity science – that are linked through their relationships. The agents can take many forms: Companies, institutions, authorities and politicians, and they can occur in everyday life on several levels.

The participants in this infinite network perform the operations, while their relationships determine the flow.

This confusing system is extremely dynamic. Participants are entering and leaving the network all the time, and their relationships change. Moreover, the system is learning and increasing its dynamics. This learning takes place both horizontally at the same organizational level – lateral learning – as well as up and down in the respective organizations – vertical learning.

Thus, the project is an extremely complex and dynamic, adaptive system in three dimensions in everyday life, which often makes it unforeseeable just a few hours ahead. But in return the network lives forever. The project stops, but the agents will be working on, now in new projects and thus in other relationships, and the project's impact remains in the system. Its learning, its losses and bankruptcies and all that happened, while the new agents that occurred are all brought forward. Everything ranges far back in time and will probably exist forever.

Our own – large, we may think – project is just a wave ripple in this universal production system, ranging from the construction of the Pyramids and the Great Chinese Wall to Toyota's ordered production of cars, and modern partisan war campaigns in Asia and the Middle East.

THE FLOW IS THE KEY: The individual participants take care of the project's operations. But their interest in their own profit often leads to sub-optimization, and thus makes the flow uncertain. The agents focus on their own effectiveness and profit rather than the total productivity or throughput, which often damages the project.

In a flow perspective, reliability is decisive, cooperation and reliable handover between participants of the results generated through the operations is the key. In a system with conflicting self-interests a more reliable flow will require cooperation, mutual learning and – not least – trust.

And where does all this lead us?

Value-Flow-Operations

Koskela's theory is the starting point for all my thoughts about the project, so it is the natural place to start as my first of three perspectives. In my adaptation, I call it the VFO theory with regard to the three key concepts: Value, Flow and Operations.

The amended order – where Koskela speaks of Transformation, Flow and Value – may sound like a detail, but it contains two crucial differences in the perception of the project. First it sets the project's sole objective – to create value – in the forefront where it should be. This value is created by the flow and the related production process, which then draws on the individual participants' operations as needed.

The operations – on which the traditional project management has its focus – thus become secondary, also in relation to the flow. How does it help that the scaffold workers can put up the scaffolding blazingly effectively if we by adjusting the flow or value slightly don't need the scaffolding at all?

Changing the word transformation – that is work generating value – to operations, which is work generating costs also brings the non-value adding tasks to light, namely inspection, transport and waiting which all contribute to the costs. Not uninteresting when we realize that more than two-thirds of the hours are spent here.

The VFO understanding thus reflects the project's eternal value-time-economy triangle with its inherent conflicts in everyday life.

While both value and operations are familiar items and included in traditional project management thinking, the flow – and the consequences of the flow thinking – is new. It is by our lack of understanding the flow and its management, that we may explain what we see as the project's strange behavior, and therefore flow is the main issue for the rest of this essay, and largely also the rest of the book. Value and operations are therefore only treated when they influence the flow thinking, such as when I discuss efficiency versus productivity.

I start by examining the phenomenon of flow and find much knowledge overlooked in our daily dealings with the project. This leads me far beyond the conventional thinking, even in lean construction. Flow appears in two different modes – either as the quiet and effective laminar flow – steady state it is also called – or as the rapid, chaotic and inefficient, turbulent flow. This is a significant phase shift, when we look at the project as a flow, where the art is to trim the flow close to the ordered, laminar side of the edge, without it tipping.

I have never really whitewater rafted, where you paddle down a wild river, but I have tried a tourist version. And my God how amusing it was and how wet we got, but at the same time I also learned something about how our calm and efficient sailing, where everyone just paddled, could suddenly tip over into chaotic situations where it seems you were just pell-mell pulled through the feisty small falls, until our guide got us back on track and into the process of bailing the boat. That day in Cody, Wyoming, I really learned something about the importance of being on the ordered side of chaos.

And it is not just about holding back, but also about understanding the other, the chaotic side.

This new and exciting world must, however, wait for the next essay, because the flow itself also provides much new to think about.

This opens my second perspective: Complex systems, because chaos only exists here. Chaos theory is a new and flourishing science that from its place of birth at the Santa Fe Institute in New Mexico, in recent years is spreading like a wildfire. This I treat in the next essay.

As the main theme of my second perspective this leads me to my third perspective of the project as co-operation and delegation, which are the keys to managing complex systems.

I elaborate further on this in the sixth essay on the autonomous project, perhaps with thoughts that may surprise and provoke, but in which I believe. They come later.

So hold on tight and fasten your seat belt, because here comes a flood of new ideas!

Flow – The Real New Perspective

Flow is the key to our community's wealth through our efficient production, trade and distribution of goods and services of all kinds. In some industries the military term logistics is used, which we also did when we experimented with Building Logistics. The strategy is in many places called lean while the IT world is talking about agile, and I have called it trimmed. Yes, a beloved child certainly has many names.

The importance of an efficient flow became apparent with Henry Ford's assembly line at the beginning of the last century, and later with Toyota's flexible car production. The project is never so simple however, because its flows have a very high variability, but flow remains the key to an increased throughput and hence the higher productivity we are aiming for.

Throughput

The core element of Project Physics is Little's Law. In 1961 the MIT physicist John D.C Little published what later became known as Little's Law. Here he simply said: Throughput equals Work in Progress divided by Cycle Time:

$$TP = \frac{WIP}{CT}$$

This requires a little further explanation. Throughput is, in my interpretation, what comes out of the factory and is sold to a customer who pays – in case of the project completed, sold and settled work. This is what we get paid for and therefore what we as a company want to increase, as Goldratt pointed out.

We can either do this by having more Work In Progress, but by doing so we are increasing our cost almost at the same rate. It is better for us in project production to reduce the Cycle Time, i.e. build faster with the same resources. And that is exactly what we do with better flow management, where we increase the share of value-added work time spent, by reducing the part that does not create value.

In other words, flow should be in the center if we want a better economy in the project while respecting the schedule.

Is it really that simple?

In principle yes. But in practice not quite as simple as that.

The Nature of Waiting

In the early 1900s, the Danish mathematician E.A Erlang (1878 - 1928) was employed at the Copenhagen Telephone Company to study waiting patterns for service to the customers of the company's central cross board. It was in the telephony childhood, when everybody with a phone had a direct line to the main central, where female telephone operators handled incoming calls by linking the caller's wire to the desired number on the big blackboard in front of her, and simultaneously sending a ringing signal.

In this simple one-to-one system these operators clearly were a bottleneck, but also – despite low wages – a cost. So how many should be on duty to ensure an adequate service quality – short waiting time – for the customers?

Erlang as a mathematician knew quite well the probability theory that was developed in the wake of earlier days' fascination with games of dice and odd bets. Erlang turned his gaze from curious questions such as: Do two in this company have birthday on the same day of the year? into waiting time.

To Erlang it was the waiting for the calls to be serviced, which was interesting, and here he developed what we today call the Queuing Theory, that mathematically explains the expected waiting time in a system with a bottleneck. Or how long on average should you expect to have to wait in the supermarket's queue when checking out with your stuffed cart Saturday morning.

Not surprisingly the waiting time depends, among other things, on how much capacity is available in proportion to the traffic, but also on how regularly arrivals take place. The smoother flow of arrivals, the shorter waiting time for the individual, or put another way, the better utilization of the capacity. We know this phenomenon when motorway works necessitate speed limits. Now the traffic flows a little slower, but more evenly, and the reduced capacity is utilized better. Often the loss of time, despite the reduced speed, is almost negligible.

In the project flow this understanding leads us to see that we can increase capacity utilization by smoothing the flow which we can do by making it more steady. This is almost always cheaper and faster than just increasing the physical capacity of crew or equipment.

I've done it myself many times, and in essence it is the approach of Last Planner, where reliability is recorded and enhanced using the indicator PPC – Percent Planned Completed. The percent of what was planned that was carried out as planned.

When we know that a third of the working time is spent on waiting, there is definitely something to be gained by increasing the reliability, and thus reducing this waiting time – as we did in the shipyard when we unidirected the traffic up and down the steep ladders. Starboard up, Port down. One of my students introduced reservation for the crane on an infill construction project in eastern Copenhagen, and increased its effective capacity quite a lot while the waiting time decreased dramatically.

Queuing theory is indeed a useful understanding, and by the way: the probability that two of the company have the same birthday is over 50% with more than twenty participants.

The Bottleneck Theory
Israel's response to Japan's Shigeo Shingo was Eliyahu Moshe Goldratt (1947 - 2011). As a successful business leader

with a background in the IT world he established himself as a consultant and built The Goldratt Institute, based on what he named The Theory of Constraints, i.e. the bottleneck theory.[2]

Alongside that he and his staff helped a large number of companies worldwide to achieve better productivity, they also wrote – and especially himself – a number of inspirational books, many told as novels, but with a serious message. He worked from the premise that it is the client himself that has to understand and realize the message, not the consultant doing it for him in everyday life.

A somewhat surprising approach from a management consultant who usually lives from the hours sold, but Goldratt preferred to get a share in the productivity gains, i.e. in the value he created. In my experience a healthy approach, for there is much to be gained in project production.

Goldratt's very simple and logical idea is that in any flow system there is one and only one bottleneck, and this determines the total throughput. Do we want to increase throughput and thereby earnings, it is this 'global' bottleneck we must open up. If we increase its capacity, inevitably a new one turns up, and then this is the one we need to concentrate on. And so we can continue in our constant quest to increase throughput.

This clear focus on the bottlenecks will naturally make critics point out that there are many other places that also can be made more efficient, and to this the only to answer is yes, because we will never reach a completely balanced process in project production. There will always be something to improve, but it should constantly be done with a focus on flow, because flow is the key to increased productivity, whereas a narrow focus on efficiency often is the very root of our problems.

Often it will be found that the bottleneck is an overloaded middle manager. We then have to either reduce his workload or give him enough help. However, this can be difficult in a traditional thinking economy such as found in construction firms, where a middle manager is only seen as a cost and not as the production prerequisite he really is. You may save on this cost as an efficiency measure, but you will destroy the flow reliability and the loss will be much greater than the savings. But to realize this requires that you understand the flow thinking, which very few do. For, as we shall see, productivity, and hence earnings, associates to flow, to throughput and to schedule in the project's eternal triangle, while efficiency focuses on operations and hence on cost.

When Efficiency Destroys Earning

In practice a dynamic flow system is never quite balanced. There will always be something waiting for something else, because all the time small and large events disturb the ideal flow – the weather, sickness, error or failure in one of the streams – and immediately a local, temporary bottleneck pops up. Almost everything is ready, but we are just waiting for the last little thing. That's what we must learn to live with.

The bottleneck will thus create a queue upstream, something waiting for the last condition coming into place and the bottleneck to dissolve – and that is a waste. As we now know from repeated studies approximately one third of the working time on construction sites is spent waiting, it is surprising that no one apparently cares about why it is so. But if they did, they would be deeply surprised because the answer is that it often is because we seek to streamline the process!

Did you expect that?

The explanation is that we are planning, contracting and working based on the traditional management thinking, where one should seek to minimize the cost of operations – the third dimension of the project's eternal triangle – which leads us to the assumption that we should seek to benefit from working 100 percent, in other words work as efficiently as possible. Altogether, everywhere and all the time. Thus we reduce the hours not creating value – for example the foremen. 'They are not building, they just cost money,' says our accountant.

And there we are. The saved foremen were the ones who were to ensure sound work and thus the decisive stable and reliable flow, but now they are so busy trying to cope with the daily problems that they have no time for the preparations. And if that was not bad enough, making full use of the system's capacity leads to bottlenecks and dramatic extensions of waiting time, which the queuing theory could have told us – if we not had forgotten or never heard about it.

But one needs not in practice to have heard of the theory, we all know the phenomenon from the motorway, where a local speed limit makes everyone go at the same speed and we get a flow with reduced variability, and the throughput increases. We exacerbate the practical capacity by changing the nature of the flow a bit.

If we look at the construction process, we urge to save by a unilateral focus on streamlining operations without understanding that we thereby create bottlenecks along with, uneven and unreliable flow. Thereby we lower the productivity because productivity is measured on the total production – the throughput – and not on the individual operations.

Our urge to save to achieve lower costs leads us in other words to destroy our flow and reduce our earnings.

It's indeed creepy!

The Bottleneck as a Control Valve

The bottleneck may from time to time be an advantage, but only if it is planned and serves a purpose. The bottleneck may be seen as a kind of control valve, which ensures a smoother flow downstream by holding things back, that will not be used right now. Thus the system downstream is not quenched, for example by materials and supplies, in cases where 'space' is the critical flow. An offsite storage can ensure this and constitute such a control valve that only lets things move ahead, that are being used within the next few hours.

Goldratt sees managing this global bottleneck as a general substitute to jumping around looking for local bottlenecks, and then adjusting the rest of the system to this one bottleneck's needs. It is reminiscent of the anthill, where all the ants adapt to the queen's and the larvae's needs. The anthill success – its throughput – is survival, and the queen is the bottleneck, while everyone else just align. And effective they are the damn ants. In tropical rain forests, the total weight of termites counts for more than half the weight of all living beings – insects, mammals, birds and reptiles.

Cooperation and division of labor pays off.

It may be a good idea to consider this strategy more systematically, especially if there is a relatively solid and visible bottleneck in the system, e.g. the dock at a shipyard. By ensuring that this bottleneck always has enough work, we maximize flow through the bottleneck by having a little more work-in-progress' than theoretically necessary. In other words, we buy us free of the effect of the inevitable variation in the flows.

In the construction of Terminal 5 at London's Heathrow airport space was a big problem, and here they handled it in a really radical way. The main contractor established two large storage areas outside the airport, and demanded that practically all deliveries took place through one of these, depending on the type of delivery. At the same time it was determined, that only materials and supplies to be used within the next 24 hours were allowed to enter the construction site itself. Combined with an effective delivery of stocks, this avoided the cramped construction site becoming 'strangled'.

While I have very effectively used the idea of a stable bottleneck in shipbuilding, where the building dock is the obvious choice, I have only rarely seen it in construction projects. But where it has been used, focus on the bottleneck worked every time. My own thinking about a better building process therefore rely heavily on Goldratt's thoughts.

Something Waiting for Something Else

If we put on our flow-glasses and look at the project as a process, we can perceive the project as something that happens almost by itself, when all the prerequisites are present.

Should we increase the process intensity, we must therefore find and regulate the critical flow – the one the bottleneck is waiting for. Here is the key to a better process: Find the critical flow and make it smoother.

I think Lauri Koskela had already realized it, but I myself was speechless when in the summer during the heat wave in 2011, this simple solution dawned on me while I sat at night wrapped in my Brazilian poncho – purchased during the Conference in 2002 – on my balcony by the kerosene lamp and speculated, while my lean brothers were at that year's conference in Lima, Peru.

If we imagine a stove, its process has three prerequisite flows, namely wood, air and heat, and one of these three is the critical one. When the wood is put into the burning chamber and stacked properly with kindle on top, it is waiting for the process heat. This comes when we light the kindling and the critical flow is now one of the other two. It is often air, if there is a heavy, cold air plug in the chimney. And when the stove is burning merrily, we can regulate by either adjusting the amount of air or the amount of wood. But more air will not help if wood is missing, and vice versa.

In construction the process is the same, it is just a little more complicated, because the process has not three but seven different groups of prerequisites:

Previous work
Space
Information

People
Equipment
Materials
External conditions

Several of these seven prerequisites, however, are not 'owned' by the project, but by the client, the authorities, the subcontractors and suppliers who also have other duties, and therefore do not always prioritize the project's best interests, but their own. Studies have also shown that there are actually many more single prerequisites, often more than fifty. In practice however, we limit them to a set of seven kinds, which makes it possible to keep track and control the flow in our everyday management.

Tangled Flow

From this understanding of flow, we meet here an infinite, but often overlooked network in the middle of the project life. Maybe we should think a bit more about the phenomenon that I call Tangled Flow.

I do not know of studies of this phenomenon within project management, if existing at all, it is probably to be found within the complexity research thinking about networks. But just understanding the seven flows that feed the process naturally leads to a new concept: Critical Flow – the one of the seven flows that slows the process at any time.

If we once again use my small example with heating the summer cabin with a wood stove, which, in order to heat, has to have three prerequisites present: Wood and oxygen – that is 'materials' – and a sufficiently high temperature, that is a result of the 'previous work'. Missing one of these three, stops the process, and so it is in the project as well. There is one and only one of the seven flows that determines the intensity, so if we want to build faster and increase throughput, we must find this and increase this critical flow.

But here we meet the tangled flow. Missing wood to burn may be due to the forest ranger being busy with other chores – lack of manpower – which weaves us into our neighbor's project – where he might wait for equipment or information to move forward. Perhaps the forest authorities' decisions delay – or maybe it's a lack of environmental approval due to a protest case, that brings it all to a halt.

Indeed a tangled flow where it ultimately may end up being our own actions that block, because we signed a protest against the new county road.

Everything is tangled up in the project's complex world, where a single critical flow can easily make a well-organized project capsize, if you are not aware of it.

The main contractor on a large shopping mall in Copenhagen a few years ago was not aware of this. In the final phase in which the great variety of shops each had to finish their retail area with their own craftsmen, space became the critical flow, but was not brought under control. This flow, and not least removal of packaging and other waste, was out of control, and the site quickly became one big battlefield, where everybody tried to secure space for themselves and constantly brought more materials from fear of running out. Chaos and delays were the result.

A few years later another main contractor solved the same problem on a similar construction project with an effective space management, that only allowed the materials that were being used the same day to come onto the site, and ensured that waste was removed immediately – and everything went just as it should.

Fluid Mechanics
When a – Danish – construction engineer talks about flow, the profession of hydraulics should naturally show up on the radar. Here we have a whole science dealing with flow. But it is usually overlooked in project management's focus on operations and budget, and I have yet to come across hydraul-

ics – or rather Fluid Mechanics – in the lean construction movement's otherwise diverse discussions of flow.

Fluid mechanics is originally about the flow of water or a little more general of liquids as a physical phenomenon. The original question in the science was therefore: How do liquids flow? Later, the thinking was extended to other media such as air and oil and also to more complex flows such as mixtures of oil and gas, which while it flows also may change between the two states depending on pressure, temperature and velocity.

In fluid mechanics we fundamentally find that flow can occur in two different ways, namely laminar or turbulent. The laminar flow is what we see in the lazy river, while we find the turbulent flow in the feisty mountain stream which often feeds the river.

Basically, these are two very different modes without a smooth transition. We speak in other words of an either-or phenomenon, although in the laminar state there are warnings of impending turbulence in small eddies, which die out and, similarly in the turbulent state of small ponds, where there is peace in the middle of chaos.

This is an interesting understanding of flow, for the most effective flow, says the theory, takes place on the laminar side but close the brink to turbulence. I will elaborate on this a little bit later.

But first the phase shift, when does it happen?

Reynolds Number

Reynolds number is probably an unfamiliar concept to most people, but quite simply put, Reynolds number is the relationship between the forces driving the flow forward and those that slow it down. The number determines when the flow jumps from the laminar to the turbulent state or – in terms of the project – when we push it into the chaotic state.

Again a critical issue in our project understanding and management, but as far as I know never investigated. It happens quite often when we have to press the project forward, and Reynolds numbers is probably worth knowing and understanding – and not least keeping an eye on! Now the project flow is not homogeneous as water or oil, but rather mixed. However, science is well aware of the hydraulics of flow of mixtures, for example the mixture of oil and gas, or solid material suspended in water, the so-called slurries known from the mining industry's transport of crushed mineral, and the timber industry's transport of pulp.

In complexity theory I have also encountered the concept of Reynolds number within economy – not least in the financial sector, where the balance on the edge of chaos is vital. The finance sector is a world, not least in the United States, that has enormous resources behind it, and much research in complexity is financed from those sources. There is money in understanding and predicting the financial market development just a few hours into the future – today even seconds – ahead of the rest of the players.

This is not where we are going in project management, but as a project manager it is also nice to understand what is happening – and to be able to interpret it and thus see a little further.

With Reynolds numbers we meet the border to a completely different world, namely that of the complex, dynamic and thus chaotic systems'.

Many might say 'so what', but it is not a simple world, for now chaos breaks loose.

The project is complex, and potentially chaotic, and therefore it requires us to rethink our whole way of looking at it.

1) Hopp, Wallace J. and Spearman, Mark I. (2000): Factory Physics, McGraw-Hill International editions, second edition

2) Goldratt showed the impact of the bottleneck in his first business novel The Goal and has dealt with the issue several times since then. In my opinion, however the novel Velocity by tree of his colleagues is maybe the most inspiring.

Goldratt, Eliyahu M. (1984): The Goal, Gower Publishing

Jacob, Dee; Bergland, Suzan and Cox, Jeff (2010): Velocity, Combining Lean, Six Sigma, and the Theory of Constraints to Achieve Breakthrough Performance. Free Press, New York

The complex project

*Where I introduce Project Physics
– and look at what non-linear systems
may teach us*

NOW THE READER MIGHT BELIEVE that I approach the non-sense. Talking about chaos and such stuff, can't be serious.

But yes, I am indeed. Flow is only the first of my three new perspectives, but with Reynolds number in mind we need to look further beyond the edge of the turbulence into the chaos that threatens on the other side. Because when the project so often ends up in this state, we must take chaos seriously and do something about it.

My work on Project Physics has again made me take up the project's complexity. This is partly due to the fact that the topic by no means is exhausted, partly that we repeatedly hear that projects are becoming increasingly complex. Therefore we should take complexity theory seriously, not least because it is a new and thriving science, which is currently used above all to explain the phenomena we encounter in everyday life. When I wrote my first papers on the project's complexity a dozen years ago, I received every week a handful of abstracts on complexity. Today this number has risen over my head and I have set up a filter giving me only one daily abstract, but always highly relevant. And when I look at the database behind I find that it up till now has had nearly a million requests.

My own writings on the subject are simultaneously downloaded and quoted more and more frequently in all parts of the world.

Here lies a treasure trove of ideas and knowledge, which should be a part of Project Physics.

The Project is a Complex System:

The project exists in an infinite network of agents, as they are called in complexity theory, connected through their relationships. Agents can take many forms: Enterprises, institutions, governments and politicians, and they can in everyday life occur on several levels: Individuals, groups, subprojects, committees, authorities and… The list is long and growing all the time, latest neighbor groups, environmental organizations and 'green frogs' have emerged with impact on particularly large civil works. These agents link the project to other projects through their participation in several projects, which in turn connect it to even more remote projects.

The participants in this infinite network perform the operations, while their relationships determine the flow.

This confusing system is extremely dynamic. Participants are entering and leaving the network all the time and their relationships change. Moreover, the system is learning and thereby increasing its dynamics. This learning takes place both horizontally at the same organizational level – lateral learning – as well as up and down in the respective organizations – vertical learning.

Thus, the project is an extremely complex system, meaning that in everyday life it is often unforeseeable just a few hours ahead.

This network lives forever. Projects may stop, but the agents are working on, now in new projects with other relationships, and the project's traces remain in the system. Its learning, its losses and bankruptcies and new agents occurred. The entire network reaches far back in time and will probably exist forever.

Our own – large, we think – project is just a ripple in this universal ocean of projects.

This opens a new approach to understanding the project: Complex systems theory, also called Chaos theory, because chaos only exists in such systems. This theory is a new and thriving science, which based in the Santa Fe Institute in New Mexico in recent years, has been spreading like wild-fire.[1]

So beware, here come some real new ideas!

Self-Organization and Emergence

Let me just make a brief summary. Newton and de Laplace brought us the order and rules and made planning possible. Poincaré shot their whole beautiful world into the ground, and Edward Lorenz reopened the non-linear world for researchers and foot soldiers like my son Rasmus and myself.

The Danish physicist Per Bak (1948 - 2002) formulated in 1987 his now recognized theory of Self Organized Criticality. It may sound a bit childish, but he studied sand piles like the ones children are making on the beach, where they let dry sand drift down to form a cone. He and his colleagues looked at how such cones grew and progressed. We as building engineers would probably say something about the slip angles, but Per Bak and his team focused on the number and size of the slides that happened in his sand piles.

Here they found that this dynamic mini system found its optimal size on its own. Small slippage occurred relatively frequently, while the greater occurred less often, and when they plotted the number and sizes into a double logarithmic coordinate system, the points formed a straight line.

This pattern was surprisingly found almost everywhere they looked at the distribution of natural phenomena, such as earthquakes or country, city and company sizes. Their hypothesis was that natural systems evolve by themselves toward their critical state, where they stay, teetering on the edge of chaos, because this is where they function optimally.

Stuart Kaufmann and his colleagues from the Santa Fe Institute took this idea a bit further and said that this is where life is born. Their thinking is based partly on studies of past living systems, partly experiments in computers with artificial life, in which terrestrial nature's development processes are simulated. [2]

A concept that also appeared in these studies is emergence – sudden phenomena that can't be foreseen by simply studying the system's individual elements.

Take for example water: By simply studying oxygen and hydrogen separately, one cannot predict phenomena like water – let alone its forms such as ice and steam, which, like wetness, waves and current are emergent phenomena.

More playfully the same thing is said about the taste of a good, well mixed, cold dry Martini.

I myself as a civil engineer use the urban scene as an example of emergence. Good cities arise not only by placing some nice houses together. Although each of them may be architectural masterpieces, they may fight each other to be the best, while the space between, where we humans move around, becomes a barren battlefield.

Urban planning was a subject of my engineering studies, and although I mostly studied traffic – the flow – something else was learned as well. Cities are emergent phenomena that occur for better or worse by themselves, if we do not tame the process.

Push or Pull

If the world was ideal and therefore Newtonian and predictable, our plans would work and the project run like a clockwork once it was launched. But the world is not so, as Poincare and Lorenz pointed out. It is not our plans that are wrong, it is what we are planning that doesn't behave as we think it should. Therefore we must control the project. This is achieved by breaking it down into smaller tasks that successively are triggered, and by correcting when the plan is not followed.

Usually projects are controlled by trying to follow a plan, which is similar to the operation of trains. The principle is called dispatch. Here a task is started when the plan says it should start. *Push* it is also called, because the tasks so to speak are pushed into execution.

But there is another method by which a task only starts if all its prerequisites are in place – and when this is the case, it starts by itself in the same way as the tree grows when all the preconditions are available. This principle is known as *Pull*.

In other words, they are two fundamentally different approaches to management, either to work by the schedule or by the situation – push or pull. In our everyday life we know of it from traffic. The traffic light is a push system. Here the traffic stops and run by what the signal and the program behind it say and the intersection stands empty even when there is no crossing traffic waiting, which generates a waste of capacity.

The roundabout on the other hand, acts as a pull system. It so to say draws traffic if there is space, and the capacity is fully exploited as long as there is traffic.

We can also say that push happens from above while pull happens from the bottom. In ordered systems – or systems which like trains require order and predictability – push management is the right approach. When we are talking about more precarious situations and flow with great variability, pull is the right approach. This suggests that we in project management should take advantage of pull control, but we nearly always use push. Engineers, lawyers and accountants usually want to see order and predictability everywhere; it is part of their nature.

But there is more to control than that; control has become a discipline in its own right, along with the development of robots and space rockets. Cybernetics it is called.

Control Theory

Without knowing the subject in depth, I have come to some conclusions in my understanding of the organization and management of a complex and dynamic system, an understanding contrary to popular opinion, which I meet in many – particularly public – projects, where management is mistaken for reporting. But management is not reporting – it is action, and action here and now, appropriate to the situation. Think of traffic. It is the person closest to the situation – often called the chauffeur – that is acting and doing it based on his best knowledge of both the situation and the objective.

The project is – like Per Bak's sand piles – rich in small and mid-size incidents, which all must be corrected, here and now, while even only a few of the big and serious ones require reporting and fine-thinking at the higher levels.

And when the whole situation – as we have seen – often is chaotic, it is the person on the spot, the last planner, who must act. Of course, not all control should be done at the absolute lowest level, but should in general take place as close to the situation as practicable. Overview may be necessary at times, but so is proximity certainly also, so we are talking about a balance in each situation.

Often I meet a tendency to kick problems upwards and to avoid liability. But in my understanding it does not hold. We should remove the fear to decide and act on our own, by making it acceptable to make mistakes. On the whole, we have much to learn about management in a complex and dynamic situation.

The right to manage at a higher level should not automatically entail a duty to control at the lower, but only the right to delegate, when balancing on the edge of chaos.

Chaos

The project rarely starts with chaos, although it may happen. On the contrary it usually starts orderly, where everything is planned, timed and organized, and from there it gradually gets more and more complicated until it all ends in ... yes exactly: Chaos!

As I mentioned in the section on flow and turbulence, chaos is not something unexpected, something that just happens, but rather a well-known and studied phenomenon, which today is subject to a whole new and rapidly growing science. Popularly it is called chaos theory and more scientifically Complex Systems Science, for it is precisely the complexity which is the focus of this science.

The seed of this new science was already found in the study of the non-linear systems' hidden oddities, which visionary mathematicians as Poincare did when he challenged de Laplace's rational thinking. Small deviations do not stay small, but on the contrary grow eerily fast, as Lorenz discovered when he switched paper in the printer.

Central to the project comprehension the chaos theory says that a plan never holds – not because it is bad but because it cannot hold!

I know I'm repeating myself, but the message is central to my understanding of the project, and I experience again and again that behind my audience's polite nods a 'Well, but anyway'.

I'm not saying that plans are useless, but I say that it is the planning that creates the usefulness, not the plan itself. In other words, it is the process and its dynamics that should be in focus, not reporting, control and supervision. This applies both to the project's everyday life, and to its planning and control.

When chaos theory has been so successful, it is due to, among other things, its ability to explain living systems of very different nature. Nations' interaction, organizations, urban growth, traffic, anthills, the origin of life, diseases, and yes ... virtually every scientific domain rears its head in the steadily growing stream of scientific publications in this field.

Now chaos is a very loose term that is even often subjective. I have therefore in my own understanding chosen to say that I perceive a system to be chaotic, when it is

Unforeseeable in the time perspective we want,
with the accuracy we need.

A simple and operational definition, which in many ways covers the problem, although it is hardly scientifically durable. But it makes chaos manageable in everyday life. We may in fact with this definition have two buttons to adjust: perspective and accuracy, and often in that way get out of the chaotic state by either shortening the timeframe or reducing the requirements for accuracy.

That is exactly what we do in Last Planner's pull-thinking.

When the Unlikely Quite Often Happens

But before I leave these philosophical considerations about theory and move towards the project, as it often unfolds, let me just mention the improbability principle. Hole in One is the golfer's wettest dream, say those of my friends who play golf. And I believe them. Hole in One is not just something that happens, it is something that is remembered. For others it is like winning 7 sans doubled in a bridge tournament, where the other tables are stranded at five of clubs, or when everything falls into place in your life and you meet the only one.

Those things happen, and ordinary sensibility confirm that it may happen once, but not twice and certainly not within weeks. But that may actually happen says the English mathematician, Professor David J. Hand in his book on the improbability principle, where he explains why the unlikely happens quite often in the real world. [3]

In fact, Per Bak said the same already in 1996 in the book How Nature Works [4] with the statement:

The probability that something unlikely will happen is very
high, because so much unlikely can happen.

In a large project, where I recently assisted, this statement was quickly changed to the term Shit Happens, said with a smile, for now everybody knew and understood that so it must be. And better: Much of the criticism of the other participants, often encountered in projects, was avoided. In complex systems, there is rarely one single sinner who is the cause of the failure, but an unfortunate coincidence of events.

By this I stop my speculations about Project Physics. Not because the subject is exhausted, it is far from it. On the contrary, a lot of relevant, but untapped knowledge is waiting for our rethinking the project and its organization and management.

Much is to be found in the natural science domains, but certainly also in the social science, because it is perhaps far more important to understand that the project is a partnership and thus an independent social system. This I will return to in my fifth essay.

The main theme of this essay leads me inevitably to my third perspective of the project as collaboration, because the key to manage complex systems is cooperation and delegation. I elaborate on that in my sixth and seventh essays about the independent and the living project, perhaps with thoughts that may surprise and provoke, but in which I believe.

Just wait and see!

1) http://www.santafe.edu/about/

2) Kauffmann, Stuart (1995): At Home in the Universe, The Search for Laws of Selforganization and Complexity. Oxford University Press

3) David J. Hand (2014): The improbability principle: why Coincidences, miracles, and rare events happen every day. Scientific American / Farrar, Straus and Giroux, New York

4) Bak, Per (1996): How nature Works – the Science of Self-Organized Criticality. Copernicus Press.

The methodical project

Where I look at how the Last Planner System works and presents an upside-down pyramid in project management

WHEN I TOLD MY FRIENDS, the three musketeers Glenn Ballard, Gregory Howell and Lauri Koskela that I would write these essays and that one of them would deal with methods and tools, several cautionary voices were heard. Careful here, do not make it a manual where the whole understanding of why is lost, they said.

I try to keep this advice in mind. But before the last two essays will continue my theoretical considerations, I would like to stop and look at how our new understanding of the project as a flow could be translated into a methodical approach to the project and its management.

It is first and foremost Last Planner – a well-proven methodology that in all its simplicity just works – I have in mind. But I will also be thinking a bit about how the project should be approached, if we really should follow the theory of its nature.

My outset is Lean Construction, a theory based approach to project management. To those not familiar with Lean Construction I have placed an essay explaining my understanding on www.theunrulyproject.com.

Let the Project Control Itself

Once you have accepted the understanding – the know why – the Last Planner System of Production Control is obviously the right approach to the project's organization and management.

Last Planner – one can't use the whole name in everyday life – is based on the understanding of the project as a flow in a complex system, which, like the roundabout, controls itself. Simple and ingenious, devised by Glenn Ballard inspired by Greg Howell – and for many today synonymous with Lean Construction.

But while lean construction is a thinking, a philosophy, an understanding and to a certain extent, a way of life, Last Planner is a method, which is a logical consequence of this way of thinking and thus to a large part the reason for lean construction in recent years spreading to all parts of world and to other industries. Last Planner works just as well in many other types of projects.

So let me try to explain Last Planner in the light of the theory, I have just presented – just to show that a good theory can really be a useful tool.

Basis for Last Planner is the understanding of the project as a complex and dynamic system, which, like all complex systems, has a built-in tendency to steer towards the brink of the chaos where it works best, but where the unforeseen and unexpected very often happens, no matter what else we do.

Last Planner therefore assumes that plans do not hold, for the simple reason that they can not hold.

Last Planner basically creates a pull-management, i.e. bottom-up governance. It is the man on the spot – the last planner – who best knows the situation, and therefore it is him – or rather them, for in construction we have many trades involved, each with their own last planner – who decides what will happen. In relation to these decisions all management above the last planners in principle only has the sole task to shut up and make sure that their men on the spot have what the last planner says that they need – here and now.

The man on the spot!

Actually, it is the same as the task the gardener has with the newly planted tree in the garden, which may need water and fertilizer, light and air and a minimum of suffocating weeds. If he ensures that the tree gets all this in the proper amounts, the tree will otherwise figure out how to grow by itself, thank you.

And preferably do it without interference.

In other words, Last Planner turns the pyramid upside down, as a long-departed Scandinavian Airline CEO also found out to do, and as General McCrystal did with the US expeditionary corps in Iraq, for that matter – him I will return to in the next essay.

A Five Step Planning

Last Planner works by a five-step planning and preparation of the operations: Should happen, Can happen, Will happen, Does happen and Has happened. It is expressed in various types of planning and preparation activities, all carried out in close cooperation between the relevant parties. I use the words planning and preparation, but in practice preparation is really the key word.

But let me explain it step by step.

SHOULD HAPPEN is what should happen if the process was ideal, but never happens in reality, because the unexpected always happens in a complex and dynamic system. Should happen is laid down in a process plan prepared by the project foremen, who in the project's everyday life operate the process with their superior superintendents responsible for the fulfilment of the project's needs.

Both must participate in the planning process because they each are key players in ensuring that the prerequisites are present, when the operation is to be executed. But observe, it is the participants that in the daily operations are going to execute the project that make the plan, and not some planners from the home office.

The Process Plan establishes the best possible process, and is the one we are aiming at. Everyone knows that the plan will not hold, but go about it anyway as a group, and shake hands on it as a deal that this is how the project should be implemented.

The Process plan thus describes what we should strive to do. The participants agree it's the best way to implement this project. It is not the project management's but the participants' own plan – and agreed upon.

Usually, it is made by all the participants simulating the project's process on 'sticky notes' in different colors – one color for each trade – where each note refers to a specific operation in the project. The notes are set up on a long – often 30 feet or more – piece of paper on a wall, where the participants so to speak negotiate how their borderlines and hand overs should be handled.

Often the plan is prepared 'backwards' starting with questions: What do we have when we have come to this state, and what are the preconditions for doing this, and in that way step by step moving backwards through the project's flow until the beginning. A true pull thinking, right from day one.

Usually a project is planned from the beginning and forwards, but doing it backwards is often much better, because thereby we find the pre-requisites we otherwise often overlook. Most of us also in our own life intuitively use this kind of planning when we have to catch the charter airline at 9:05 at the airport and have to close down the house and take the dog to the kennel before. We count down.

Or as the world famous Danish philosopher Søren Kierkegaard (1813 - 1855) expressed it in his book Philosophical Crumbles (1844):

Life is understood backwards, but must be lived forwards.

So in his spirit we try to compensate for that by at least planning backwards.

Among the other – large – benefits of the process planning is that all have to consider the project and their coming cooperation. The process plan is not about how we undertake our individual operations, but a look upon the project's flow, i.e. how we co-operate and hand over the completed tasks. All are standing by the wall, all talk with each other – and suddenly we have created a team.

Unfeasible solutions in the project are being discovered in time – the project's design manager is of course participating and may decide on unclear details and fix them. Components with long lead time, approvals or the client's decisions are identified and put into the To Do list, which is the key to Can happen.

CAN HAPPEN is where the tasks in the coming weeks are made ready. Here the superintendents concentrate their thinking jointly on looking into the near future, say 3-5 weeks ahead, and maybe beyond, to ensure that everything that should be used in that period really will be in place, and the tasks will be sound when they come up for execution.

Here, in other words, it is the logistics and the seven prerequisites' flow that are in focus, not the tasks themselves.

WILL HAPPEN is the last planners' agreement on the next week's tasks. It is here they shake hands – this is what we will do in the week to come.

The Weekly Plan looks primarily at the work of the next week day by day, but also looks at the following week to ensure that there are sound tasks ahead.

HAPPENS is unfolded on a daily basis by a – quite short – morning meeting, where the group of last planners stick their heads together and in detail co-ordinate the day's tasks.

And finally **HAPPENED** – which is the crucial learning which is a major issue in every complex system's development.

Errors are too expensive to be hidden, they are needed for learning. Here focus is put on what did still not happen as foreseen, the reasons are found and are removed as soon as possible. The indicator PPC – Percentage of the Planned that was Completed – is an important tool because it tells us how reliable we were in our flow.

The way PPC is calculated says much about the flow-thinking behind the Last Planner. PPC is calculated just by looking at how many of the tasks that were fully completed as planned out of the total number of tasks in the plan for the same period. Not something about almost, but quite complete, ready to hand over to the next in the chain. And not about the amount of work been done, but only about the reliability in the flow.

Just like Goldratt says in his work: *The strength of the chain is not a question about its weight, but about its weakest link.*

Where else does this happen in today's economy fixated project management?

On the surface Last Planner looks like a method, but it is basically a completely new thinking. The project is teetering on the edge of chaos, where the unlikely quite often happens. Therefore it is the man on the spot who should act, and this leads to a pull logistics. And it requires at the same time high reliability and therefore cooperation.

All the pieces seem to fall into place – and we have a solid foundation for a new kind of project management.

When Greg Howell and I – an evening fifteen years ago, at his home in Idaho after a long and thoughtful walk with Sonja and his dog Nanna up the valley behind his home – were comparing Last Planner with Building Logistics, we came to the astonishing conclusion, that whether you controlled the tasks or the materials supply, you would find a method like Last Planner. And the next morning Greg came down to breakfast with a small, closely written Post-It note and said he had thought about it and had reached the idea, that the method would turn up in managing any of the seven flows.

Hello!

Last Planner is a generic method that pops up by itself, if you think flow. And a good theory is indeed the most practical tool you can wish for.

So there's light ahead.

It Works – But Only with the Understanding on Board

Last Planner works if the method is used correctly. From all over the world reports come on how well it works. Glenn Ballard travels the globe talking, and everywhere consultants pop up, offering help to introduce the method.

So it's a great success?

But no, not quite, because Know Why is rarely included. Some smart young engineers may have put the method into a spreadsheet or another IT system; others have formalized the forms and routines, and thus embraced the original simple and ingenious idea that it is the skilled man with the rubber boots down in the hole, who know most about the situation and what he needs.

Here and now, please!

I myself have over the years gone in the opposite direction by making the last planner message as simple as possible, often down to the statement:

Ascertain that things can happen, when they shall happen, and avoid repeating your mistakes.

I can clearly see that there is not much fee for a consultant paid by the hour in saying it that simple – just fourteen words – but that is the truth, and in my experience usually it is accepted immediately among the crews on the construction site, at the shipyard or by the programmers in the IT-office, while it takes a little longer for the middle management staff to catch the message. They just don't think that it can be that simple, but indeed it is if you just learn to think flow – but that is often difficult, and therefore there is still a job for the consultant as a coach in this process of changing the thinking.

When I tell experienced project managers about this simple method, I hear quite frequently the statement: 'It is almost the same as what we already do!'

But it is almost never so when I ask in more detail. Almost every time it turns up that they are not thinking flow, but operations. The flow-thinking is indeed very distant from today's rational western thinking.

Some years ago, I became a little tired of these discussions and wrote a checklist which today I put into the hand of all those who believe that they already do it. It has helped, but not quite enough, and it's very seldom that I meet somebody who really takes the message to heart.[1]

But how to do it in practice I hear someone saying?

Yeah, well and boom, boom as a leading member of the Danish Parliament so wisely responded when asked what he would do after an election defeat, for what else is there to say.

But anyway, let me try to reflect through the project. Naturally enough, I would start from theory and look at the project through the filters Value, Flow and Operations.

First, I would want to find out what the project is all about. What problem is the project supposed to solve, and what is the right approach? Is it a space problem, is the solution then to build, or to rent or to buy, or to make the organization lean, or to outsource or…? My senior partner J.K Nielsen took me by surprise, when he before starting a fairly large project to expand the airport transit hotel in Kangerlussuaq in Greenland, suggested that a cheaper and better solution might be to buy an extra helicopter or two to ferry the travelers faster to their final destinations.

Our client followed the advice and we lost a job, but we gained at the end, because the client trusted our advice and numerous projects came streaming our way.

There are indeed often many different approaches to solving a problem, and much too often one of these is chosen without deeper considerations.

But in my hypothetical project I would then more deeply consider the process. Should the problem be solved by building and if so, then with a turn-key contractor, a general contractor or a bunch of trade contractors? Or would it be better to leave the building process to a developer or a professional investor? And so I would continue until all was a clarified and written down as a basis for the decision on the project's execution, which all participants in the further process should know and understand.

With this in place, I would start looking for my project participants.

Choose your Partners Carefully

As in so many things in life, choosing our partners in the project carefully applies here. And here it is even more difficult, because we must usually choose several partners for our project. And in doing so we can't be too careful, because we soon will have to act as a team of players, all playing together in a very fast game like basketball.

But in most projects we start by shooting ourselves in the foot, because we choose the participants through a competition.

Avoid the Competition's Curse

A construction project will typically be kicked off with a competition, not least if it is a public project, where competition at least in Europe is almost mandatory by laws and regulation. But what is its rationale – and does that hold in today's reality?

The competition is based on the assumption that the client knows what he wants, and that he is able to express his wishes in the program in a manner which participants can understand and be able to translate into their own visions, formulated in a form understandable to the client. Indeed, a very complicated dialogue to be carried out in a very formal way, and in today's Europe over the language barriers within the European community.

And here we encounter the first fundamental mistake, which stems from our rational world view. The client is almost never qualified to formulate his demands to something non-existent. His wishes must be formulated in a creative dialogue as part of the process itself, wherein thoughts and visions are exchanged face to face.

The competition is the sure way to the wrong track, if we want a sound project, where economy and time are respected just as quality should be.

One of our generation's leading Danish architects, Boye Lundgaard (1943 – 2004), put it this way, when I once discussed the competitions with him: Few clients realize how much they lock a project when asking us just to make a few sketches.

It is probably also only public and semi-public clients – and perhaps amateurs – who uphold this primitive form for the selection of partners, and perhaps only – in the case of the public ones – because legislation pushes them into it. If they really understood the project's nature not even many accountants and lawyers would hardly get far with the requirement to choose the team by lowest price and not by trust.

I have myself tried to argue the selection of a main contractor by trust and 'reasonable cost' to our residents' association, and I must say that it is not easy, but that it can be done. As a matter of fact, sometimes I feel that a competition is a convenient excuse to avoid a time-consuming dialogue.

But let me move on to the project itself.

The key to my understanding of the new project and its management will be Value, Flow and Operations. Last Planner will naturally remain high in my thinking, but first I would try to understand the project from the three perspectives, and then organize it accordingly.

Value

To create value is the whole purpose of the project, and this perspective must therefore be a key element in the management of any project, also outside the building industry. But alas, this aspect is cruelly mismanaged in much of the project management that I have seen in my professional life, where I myself have been professionally brought up with the basic assumption that the client's interests were everything.

First, you deliver your best advice and your full work power and all of your attention, until his problem is solved, and then you expect that he honors your efforts appropriately, so that both he and you are happy and can meet again for the challenges of the next project.

You think of course of the client's project all the time, both while you work and when you walk the dog, sail your boat or play with your pals. The client's project and possible better solutions next time are always on your mind, and you support his own process with suggestions, that he may reject or embrace as his own. Not least if he uses them to obtain new projects for you both to enjoy.

Cooperation with your client and your team should be like a team game if it is to succeed, and trust is the keyword. Mind you, confidence must be everywhere, and that includes both your own organization and the client's.

The creation of value is based on good cooperation and mutual benefit and understanding, that is in itself a value in the project. In my younger years as a consultant we had relationships with our clients that were like a friendship, and we knew and understood each other and created amazing results through this synergy.

The value is in fact difficult to quantify, while the cost seems to be quite exact. Often you therefore tend to use price as a criterion for the selection of your partners instead of value. But the price is not at all accurate in practice, as projects without claims and extra invoices are almost an utopia in the real world.

To put it into the eternal triangle's perspective: Concern for operations and money reduces the value and damages almost certainly any hope of a good cooperation and an efficient flow – and thereby destroys the productivity.

A Common Creative Process

Should I myself as a client start a new project, my first challenge would be to choose who I want on the team?

We are together to 'win', and I am not building to save money, although the budget must be kept. However, so should the schedule, and indeed my whole group of stakeholders should be happy, all the way through until the building is one day torn down again.

This is how I would define my rules for the game – and then welcome 'applications' to participation, if I could just be allowed – as a team manager like a football coach – to put my team together. No tendering, but just application, the same as I would do if I was looking for a new employee in my own business.

I would interview the possible participants, and not their companies' managers, but the key employees they suggest for my project. And I would choose those I related best to – in the expectation that they would be the people best able to convey my visions back to their own hinterland, in the same way as I would have to do in my own system. Indeed, we have a vital bottleneck in the flow of information at this location. The same would apply when choosing partners for the execution. Flow would be my concern, and efficient flow requires reliability and trust, and therefore an easy and friendly form of cooperation.

Then I would start a series of workshops, where we all could get to know each other and find our way of cooperation – or discover in time that we do not match. Many times our problems in cooperating are that we do not think in the same way and do not speak the same language, so we need to meet and be sure we can work together, before the project can find its way to success in a harmonious family. We create only the environment and establish the prerequisites, while it is the project itself as a process that generates its success or failure. My best projects have always been those in which all participants have been pleased with the progress, and fortunately over the years there have been a quite a lot of those.

There are many ways to start a good cooperation – the same way as a good friendship, and that is what is needed in the project. I have experienced some 'consultants' exercises, but they rarely have convinced me that this is the right way, but it's all a matter of taste. I myself prefer to sit down and friendly discuss the project's theory and daily life in a relaxed form and tell the participants that we are here to work, and that if we do it properly, there will be enough money in the budget to make sure we can all go home happy.

Value Management
With the VFO theory in mind, it is quite natural that I organize my project with three deputy project managers – one for each of the three perspectives in the eternal triangle, and each with his own clear tasks in relation to the VFO theory.

So first of all: Where is today's project's value manager?
When the project's objective is to create value, it is only natural that there is a person in the project management to ensure that this happens. Every day and at all times from the outset to the very ending. That the project delivers the expected value, and that its associated process also does.
But the value is already determined through the building

program or the basic design, some will say. But what about the process, I ask. Does the client sleep quietly at night trusting that his project will finish as planned? Has the surrounding world confidence for the project to deliver what we have promised?

Value management is perhaps not the largest job in my project management once the project has started, but it is extremely important, because to create value is the project's sole purpose.

Therefore a diligent value manager will be a key person in my project. The task is not necessarily large, but it may have a tremendous impact on the success of the project. Behind the client there are a wide range of other stakeholders and there must be someone in my organization who has the time to listen to them, and bring their concerns into the process every day.

Unfortunately, I have never had the opportunity in practice to try such a value management, so I can only speculate on what the value manager should take care of. Should I do it today, I would hire a person with strong social skills, one that could create a feeling of confidence and happiness, but in a very serious way.

In the Danish natural gas transmission project, we had a group we called Systems Engineering. Here a small group of engineers watched the very dynamic project constantly and made sure that it met the client's technical expectations. Very useful, but that was only part of the value management, important as it ever was.

Another element was that our own home was always open and every opportunity was used for a party, often at NIRAS' expense. We had many foreign participants in the project, and getting them out of their hotel rooms and into our big country kitchen, laid the foundation for a good cooperation. Many problems were solved by the kitchen table.

But did it contribute to the creation of value, I hear my critical editor muttering in the background, and yes, it did. We found many stumbling blocks by the open fireplace in the kitchen and removed them immediately. Often, the client was also a houseguest and what Greg Howell calls the Two Beer Questions were not only answered in such evenings, but brought us even closer to the project's success.

Later, when I worked with the domestic gas supply projects, I wrote a little epistle titled: What does a natural gas company do? It was in the middle of the hectic construction phase, but my answer to my own odd question was that it supplies energy. Not gas, but just energy, something all the future customers already had in the form of oil-fired burners, stoves or whatever. So, I wrote, you have to go out and sell a value that the customers already have. The Danish legislation reduced the opportunities for a genuine price competition, so there were obviously few other benefits to sell. These it was their sales people's challenge to find.

But, as I wrote, you can destroy it all in the construction phase, if your piping contractors are acting brutally, as construction contractors often are. Have they just left one front yard as a battlefield, you can wave goodbye to more customers on that road or in the homeowners' association. In other words, don't choose your contractors by the lowest price, but by the best value. And this applies not just to gas lines, but just as much to rehabilitation projects where we work in people's own home.

And it is also the same in our community, where we often create an unaccountable number of excavations for tunnels and roads, just because this is the cheapest option.

When the Øresund link between Denmark and Sweden was built, the client made quite an effort to keep the neighbors happy, a value often forgotten in major construction projects. His criteria was defined and written down and their fulfilment were measured regularly as the project progressed.

Actually, my value manager will get quite a few tasks on her table throughout my project

Flow

In 1997, Glenn Ballard wrote a paper called: Look Ahead Planning – the Missing Link in Lean Construction. Here he pointed to the need to look ahead in the project, and to prepare the tasks that would emerge in the coming weeks. In other words, the logistics. This is what happens in Last Planner's Can happen planning.[2]

Glenn made the preparation of the project's operations into a discipline in itself, and moved the focus of management from the operations to the flow, and further to the preparation, and thus to establish reliability. Factory Physics had opened up the queuing theory to him and to Greg Howell, and they realized that reliability is the key to a better flow and thereby to increase the practical system capacity, through a faster process with lower costs.

The increased reliability transformed parts of the two-thirds of the working hours that are not creating value, into value-adding work, all by itself. With better logistics the work just runs smoother. The project runs faster, the craftsman working on piece rates earn better, because their slack time is reduced, and everybody is happy.

Flow Management

In our tests with Building Logistics we realized very early that a person would be needed to take care of the logistics. For lack of a better name we called him the provider, a name at that time used for the agent that made sure that everything was in order for the silver anniversary in the village hall. A fine concept, and with the right woman or man in the job we saw it work again and again. First we saw the provider as a supplement to the construction manager, who could then concentrate on formal issues such as extra work, quality assurance and payment. The provider's job, on the other hand, was to coordinate and to make sure that the entire flow worked.

And it worked! In fact so well, that to everyone's amazement we could almost do without the construction manager.

Today, a dedicated Process Manager would therefore be a key person in my project. Her primary task would be to make Last Planner operate in our everyday work and to facilitate cooperation.

A few forward-thinking Danish contractors have already seen this, and have moved progress meetings away from the construction site and back home to the head office. If the crews themselves manage the daily flow, with the process manager as a coordinator, they should not be disturbed by the progress meeting's formalities, money talk and discussions, that can easily destroy the good atmosphere in the project's everyday life.

Operations

The operations are so to speak the project's engine room. It is here the work is done with yellow machines and men with hard hats and safety footwear. That is how we all understand construction. It is also here the money is spent, and it is here we would like to see the well-oiled, efficient process. But it is also where we often commit the most serious errors in organizing the project.

We try to save money!

The competition pushes the price down and no one seems quite to understand that it is the project's job to spend money, albeit do it sensibly. But it has nothing to do with saving money. Once we have ascertained that the value to be generated is worth the costs, the quest is to realize this value, which we have found reasonable. We should then set out with the knowledge that this project will cost us X million, and so it is. The objective is then to ensure that the value is achieved and that it is done within the budget and the schedule. The time is taken care of by managing the flow, which at the same time facilitates the costs by reducing waste, especially waste of time.

In other words, we should aim for the best process and the most reliable flow, rather than the cheapest! Every time and all the time, because at the very end it is the most productive process that gives us the lowest costs. If we can't afford to build, we should rather abstain from doing so in the first place, instead of mismanaging a project trying to spend the least.

Unfortunately, our practice sets us up to select our participants by lowest price, and thus we buy our project failures ourselves.

Lowest price leads inevitably to unreliability!

The bidders are always looking for loopholes in the tendering documents, through which they can later milk extra costs, and they win by a price that in any case is going to stress their own production capacity to the extreme – and therefore triggers the delays and unreliability, which destroy the flow and the other participants' business. This again leads to new claims and fights and bad atmosphere all the way around, where the goal should be cooperation and efficient flow and a happy day for all.

Ah yes, we buy our own problems in the project, and we pay dearly for them!

Therefore my approach is, assuming it is possible, to choose my participants by reliability. Is there a need for a fixed price, I will ask them to offer it and let it be checked by an independent quantity surveyor – but not by an alternative price, for an alternative price shows distrust – and then make a deal if the price is fair. If the contractor's and the surveyor's prices do not match, there is usually an error in the assessment somewhere, and the solution is to find the error before the deal is signed.

We do not, as I have already said, build to save money, and at least not on the faults of the other parties, and the project's success lies in trust, reliability, cooperation and fair business for all. Nobody benefits by being participants in a project where somebody loses money because they have calculated wrongly in their tender.

Here I would allocate an appropriate budget reserve, because the unexpected happens and happens quite often in a project. When it happens, I would again be approachable when it came to obvious calculation errors or real unforeseeable events. I'm not undertaking the project to save, but rather to spend money to create value, and to keep up the good cooperation and the smooth flow.

Right up to the budget, that is, but definitely no longer, because that is where I begin loosing value. If I spend more than my budget it will influence other activities in my group of stakeholders.

It is rare that I have seen this apparent asymmetry in the price expressed clearly, but it has happened. A penny saved within the budget rarely has the same value as the pain of one spend above the budget.

Contract Management

Contract Management I will call the leadership of the triangle's third dimension: The Operations. For that is what is left as the third leg of the triangle, when the schedule is taken hand of by the flow management, as we may call the last planners with their rubber boots. But this little annoying task, which easily overshadows the others in the usual approach to project management, must by its nature be taken care of. My own and others' experience is that this task becomes much easier, if the other two targets value and flow are taken care of. But one can never be completely safe, even if one has chosen the participants by confidence and not by price.

There must be order in the agreements, the quality assurance and the budget.

The Wholeness

But who takes care of the whole, someone will probably ask. By nature, it is the project manager himself, is my simple answer. And he does it not least by immediately starting to map the project and its execution with his three deputy project managers and defining the duties each will undertake in each phase of the project's life cycle. And at the same time choosing the methods and tools that seem relevant.

It all comes together in an overall plan of action, that is fundamentally aimed at the daily work. The plan is introduced to all participants, and is updated at least at each phase shift, adjusted in the light of experience. It is also here that it is decided which indicators to use in monitoring the process and its reliability. PPC is almost mandatory, but there are many other, each of which also says something about the process' own soundness.

I have myself successfully used 'order on the construction site', 'number of rush orders', 'amount of sick leaves' and 'number of project clarifications' as expressions for the various flows' reliability. On my own list there are more than 30 to choose from, and it is still growing. Correspondingly there are numerous tools that can be used as needed, but only a few at the same time, please. Or better yet, let the participants choose additional indicators, but no more than four besides PPC.

However, you can replace these four at the status meetings being held with the whole team every three months.

Looking closer at the Last Planner, you will find the common binding process planning – the Post-It planning – as something central. This tool can be used almost anywhere there are plans agreed to.

Another tool I have successfully used is the instant minutes of meetings. An assistant or a secretary attends the meeting and writes immediately the minutes, which are shown on the big screen so that everyone can follow what is written and, if necessary raise objections here and now. This, so to speak forces decisions onto the system and there cannot later be objections to what was previously agreed. The minutes are mailed to all participants immediately after the meeting, and is in other words a reliable agreement.

The method may surprise at first, but with adequate co-operation all will soon see the benefit. And peer pressure will often silence the troublemaker who from time to time has joined the meeting.

There are indeed plenty of tools, so one must be careful

not to assemble too many from the rich buffet.

And reliability should always be at the center, for it is here that the project very often is derailed.

---

1) http://www.theunrulyproject.com

2) Ballard, Glenn (1997): Lookahead Planning: The Missing Link in Production Control, IGLC-5, Gold Cost, Australia.

The autonomous project

Where I look at what the social sciences may teach us. And where the art of war enters the story

UNDERSTANDING THE PROJECT as a complex system on the edge of chaos leads to a completely different approach to its organization and management. Phenomena such as unforeseen and improbable events, emergence and chaos are to be expected in the complex project, and they require the ability to act quickly and often by yourself or in small groups, groups that often occur in the situation and fit badly into a top-down project management with rigid plans and procedures.

Modern warfare has, from Vietnam and onwards, seen the surprising force with which small, intimate, self-employed groups of partisans have been able to fight against well-organized, well-armed and on the paper far superior conventional units under a central command. That's what General McCrystal realized in Iraq with his Team of Teams strategy.

The generals have learned that the plans themselves are nothing, planning is everything. The value of the plan is that it is made, and that the task ahead is worked through, but once the campaign starts in earnest, the plan is only a description of what we ought to do. What we will do depends naturally on what we can do in the current situation here and now and on the spot.

It will in other words be a pull process in which it is the situation here and now that determines what is going to happen, and a situation where the process itself must draw the necessary prerequisites into place. A pull logistics, in contrast to the conventional push logistics that governs rigidly, based on what the plan says.

In a complex system in which the plans do not hold because they cannot hold, a push management is obviously inappropriate, because too many impossible actions inevitably will be launched. The history of war is rich in such disasters.

Pull control requires on the other hand, that the man on the spot is reliable and able to assess the situation, including the need for more prerequisites to be in place before the action can be started.

The man on the spot – the last planner – suddenly becomes the key person in the management of the complex project, and everyone else's role will be to support these last planners with the prerequisites they need – information, materials, people, equipment, etc – to ensure sound operations. But it is the last planner who says OK and put the assignment into the plan for next week, and removes it again at the daily morning meeting, if it is not sound after all.

Last Planner – Once Again

I have repeatedly referred to, and just explained the Last Planner system, which is rightly regarded as one of the key elements of Lean Construction. Last Planner System is a brilliant translation of the complex project's need of a pull control of the flow to the project's everyday practice, and it works. Therefore I use the method again as my approach to what I call the self-organized project.

When I met Last Planner for the first time, it was not really something new for me. We had worked partly with the same principles in *Building Logistics*, and I also knew the whole thinking from my many projects in Greenland, where the challenge had been the same: Plans do not hold, the unexpected happens, so let the man on the site assess the situation and act.

It required that we have chosen the right man and he made the right decision and not made too many errors, but that was a condition accepted throughout the system. Later we got better and not least cheaper communication, and the confidence in the man on the spot became reduced in importance, while the reporting requirements grew. I'm not sure this progress was for the better, because the remote control began to take over. But complex systems have to be managed at the situation, if the process is to be maintained laminar, and the flow not jumping into the turbulent, chaotic state.

However, it is only as I write these essays, that it really dawns on me that the Last Planner is not a method or a tool, but basically a way of life. But really, the method is logical in the project's unpredictable universe. It is the man on the spot who has the best perception of the situation; maybe he lacks some information, but then we should rather support him for a better overview and allow him to act, than try to act without detailed knowledge of the situation on the ground.

It was precisely what General McCrystal realized in his 'team of teams' strategy and what, under the Department of Defense's Command and Control Research Program, has been described by David S. Alberts and Richard E. Hyes in their book *Power to the Edge, Command… Control… in the Information Age.*[1]

It is the process itself that acts – as the tree grows – and as leaders we can only regulate the environment and manage the logistics, the flow of prerequisites. In the case of the tree, we are talking about water, fertilizer, space, light and air, while we in the project talk about the seven flows: Previous work, space, information, manpower, equipment, materials and external conditions. Only when the all prerequisites are in place, the tree will grow or the project progress.

Let me with this in mind repeat the method behind Last Planner, but this time focusing on delegation and thereby on the autonomous project.

The Man on the Spot

It is in everyday life it all happens. This is where people create value by building houses or ships or IT-systems, or whatever the outcome of the project is, and this is where independence enters for the first time. At least in the Danish building process where the craftsman knows very well how a job is performed, and almost all the artisans that I have met also have been interested in doing a good job. Many have even been proud of what they handed over to me.

To delegate is to show confidence and, in my experience it is often rewarded with an additional interest in the job. Now it is not to get something done because someone has said to do so, for example, to dig this stupid hole, now it is the job of installing this pit and connect it to the sewer, so the water can be pumped away. The task makes sense and gives the man on the spot a responsibility, because the next guy is waiting to take over the finished work in the further flow of work.

He is not just 'arms and legs'; he is an employee who has his place in the team, but now also in the rest of the system. His professional competence comes into play and is frequently applied by himself. I've seen it plenty of times when in everyday life I took the time to explain why and then leave how to my staff.

My secretary many years ago still remembers when we sometimes meet in the shopping mall, how I entrusted the entire operation of our 'front office' – our typing, filing, shopping and all the other practical issues – to her with a few words about what was important.

And I remember it at the same time as something that just seemed to work. We got reports and other writing out in good order and former documents could always be found – without IT, but with the primitive systems we gradually introduced.

I may be wrong, but no one seems to understand this in the building process. They hire the cheapest labor instead of the best. Perhaps the man is cheap per hour when he just digs, but when the hole is about to crash down, or the task is otherwise dangerous and he just continues digging, the problems arise and it is not at all cheap anymore.

Always find the right man and make him an employee instead of a worker. Involve him, give him responsibility and stimulate him to make suggestions for improvements which you obviously receive in a constructive way. Kai Zen says the management terminology with a fine word from Toyota, but in plain language and with the rubber boots on, it sounds from me: What do you think?

Then we are at the same level in the hole, and Karl is suddenly allowed to express his opinion on all this shit, without cleaning his boots and set them in the front room and sneak into the office on his stocking feet with his helmet in hand.

Karl has become a member of the team!

The Weekly Work Plan

In the project it is thus the man on the spot who himself assesses the situation, and he does not include a task in the weekly work plan, if the task is not sound, that is, if not all seven prerequisites will be in order at the expected time of execution. An unsound task is unreliable and may therefore disturb the flow of the process, precisely where reliability is the central issue. That is, the reliability of the handover of the finished results of each task to the next link in the chain.

The weekly plan is agreed at gang level each week, and is the agreement for the next week's work, but it also looks an additional week ahead, in order to identify upcoming tasks and make certain that they will be sound and to ensure that the there is something to do for all the participants.

In the best of my projects, I have found that two or three gang leaders came to the meeting with a plan for the floor where they worked and said: We plan to do so and so next week, and then this part of the plan was in place. What else than OK, could I say?

It is here the *will* element comes in.

The weekly plan is checked in a short meeting every morning where today's tasks are confirmed and practical details agreed.

The plan's reliability is established by the very simple indicator PPC – Percentage of the Planned tasks which were Completed. PPC is thus an indicator of how many of the tasks that should have been completed, actually were. Because PPC only looks at the reliability of the flow, there are only two answers, Yes or No. And with a Yes the task was completely finished and approved by the teams to take over the outcome. The space was cleared up, excess and waste materials removed and in every way everything was ready for the next task to begin undeterred. PPC expresses reliability, and a task not finished points to a weak part of the flow, especially if there is a failure in the same place again and again. Here interference is immediate, increasing reliability throughout the project. Usually I seek not 100% because it can be obtained just by under loading the capacity, but rather 80-90%, as the weak links thereby are revealed and can be strengthened.

PPC is therefore assessing what actually happened in order to get the right cooperation established, where it often becomes a sport to comply with the plan.

In a rehabilitation project in Vesterbro, Copenhagen we had arranged a kind of topping-out celebration on a Friday afternoon with beer and burgers. Almost everybody were present, but not the painters and I wondered.

I climbed up to the fifth floor, where all the men stood painting for their dear life. I reminded them that we were sitting down in the shed and missed them. 'No,' they said, 'we have all said in the weekly work plan that we should finish these apartments today, and so we must. Even my arguments that there was certainly not anyone who should take over from them on Monday morning, could lure them down to the beer.

They had promised something and they kept their promise, period!

The Look Ahead Plan

While the daily and weekly plans are usually fairly simple to get going, the Look Ahead or 3-week plan is often more difficult, for here we move into 'sacred' territory.

The Last Planners take over much of the daily work planning from the often busy and stressed superintendents. These good people have thus free time to use making future assignments sound, typically three to five weeks ahead of time.

But they often instead feel redundant. Usually they have been almost high in their hour to hour 'firefighting', where they were running around with the phone in their ear and bashing with arms and legs; and now suddenly things are quiet all around and the project grows just by itself. They are becoming gardeners instead of being warriors.

But don't worry, there is work enough to do, but now it is no more the operations, but the flow and the logistics that should be the key issue for them.

I remember the shock I had at a meeting with the Executive Board at a large shipyard when I presented the concept, and the Economy Director immediately found that by that they could save 25 middle managers.

It took me an extra cup of coffee and three deep breaths before I had pulled myself together and could say that he in no way could do that. These experienced people were not redundant, they should just learn to look ahead in the process and create reliability.

It can't be right that three days before the agreed delivery date of an auxiliary engine for your ship, you are told that unfortunately it is three weeks delayed, I said. Better to use these experienced foremen to keep track of your suppliers' reliability and let them possibly even help these suppliers to become more reliable and you will save a lot in your process. That's what Toyota does.

There was a strange silence around the table and I felt, that in a few minutes my work as a consultant would be terminated. Should we help our suppliers to increase their own productivity, they asked, and I said: Yes, if it increases your own reliability and thus your throughput!

In everyday construction, the weekly look ahead meetings create this reliability. This is where all the superintendents or whatever they are called nowadays, agree on who takes care of each of the seven conditions for each upcoming assignment. Here too PPC is used to increase reliability, but a reliable flow of all prerequisites is mandatory, so here we seek 100% PPC.

It is also at this meeting the process is evaluated as a whole. Are we complying with the overall schedule, and if we are falling behind, in which of the seven flows is the constraint?

It's just as when the tree will not grow, where the challenge is to find the missing prerequisite and fix it. If it in the project is the flow of information that is the problem, it does not help having more crew brought in, which is often the action chosen. But doing this just increases the costs and sometimes creates further problems by eating up space, which might be the next constraint.

The Look Ahead Planning is therefore an extremely important feature of Last Planner, now made based on a solid understanding of the project as a process in which control of flow is highly important.

The Look Ahead plan thus expresses the can do element.

For the participating superintendents several things often happen. Firstly, they become more relaxed and get more time for their family and the dog, and secondly, they learn to look ahead and anticipate the problems that may turn up.

The time horizon of this planning depends of course on the nature of the project, and for years I thought also of its size. Remodeling of a bathroom: 2 weeks, a new carport with a storage room: 3 weeks, a new center for elderly citizens 4 weeks and a super hospital or a concert hall: 5 weeks. Doesn't that sound logical?

But today I do not think that this is so. The project may be greater, but it may not require a longer time frame in its logistics. There may be flows with long lead time, and they must be followed separately, but I am more and more convinced that three or a maximum of four weeks should be the rule with supply chains as we find in Denmark.

The Process Plan

This whole execution of the project can obviously not take place without an overall plan. It's what we in Denmark call the process plan, because it maps the best process for all parties. In other countries it has other names, but in my understanding mapping the best process is its key purpose.

The plan is made at the start of each phase, in cooperation with the participating foremen and superintendents. Here, by working together they walk through the expected process and find the best way through, with due consideration to all interests and with respect for the master plan's mile-stones.

The process plan is frequently made by a pull planning, where the planning starts with the end result and gradually works its way backwards through the prerequisites for the tasks identified. It happens in a session as a Post-It planning, where the parties together play out the logistics by sticking Post-It notes, each representing a task – each trade its own color – up in the best sequence on a wall or a roll of paper, clarifying and negotiating their mutual dependencies.

The Process plan expresses thus the should element in the Last Planner.

Again, the delegation of responsibility pops up. It is the contractors themselves that are agreeing on the best process, not the project management. Maybe it is not the very best process that is agreed – that we rarely know until afterwards – but if Danish craftsmen have discussed and given a handshake to a specific process, then the process indeed is also the best, and it is what is usually delivered.

And more importantly, they own from now on the process themselves and can rightly suggest improvements, which almost always happens.

Believe it or not, job satisfaction increases and the number of accidents drops significantly, and so the success is certain. There will in earnest be created reliability in the flow.

Reliability

The project process is the result of a complex and dynamic flow with the risk of turbulence and thus chaos. A detailed control top down is beyond our opportunities in practice, and as leaders we have only the possibility to create reliability, disseminate information, delegate and build on trust.

Stalin said that trust is good but control is better. The Danish professor of political science – and trust – Gert Tinggaard Svendsen has turned it around and says control is good, but trust is cheaper. Trust is the hidden source of the wealth of the Scandinavian Tribes, he argues, for confidence is generating more confidence, initiative, resourcefulness and willingness to provide. Productivity and the quality of the work becomes higher when a company shows trust in its employees. People who are shown trust simply perform better and are more satisfied.

So here we have a free resource that is just waiting to be released.

Trust is also a capital that can be drawn upon when the unexpected events choose to occur. And they occur because in the production's economy, lies an incentive to increase the velocity, which is pushing the process to the edge of chaos, where even small disruptions can trigger the fatal transition from laminar to turbulent flow and chaos.

The project's eternal triangle, the struggle between value, time and costs contains latent risks of chaos. So it must be when all three dimensions must exist in balance, and our handling of this situation should respect the risks through trust and reliability.

However, today's treatment of the three dimensions of the project life is not based on creating trust, but rather on expectation of conflict. We choose participants either by lowest cost or highest beauty, just as in past marriages where it was the largest dowry or most prestige that spoke, more than quiet love. We throw ourselves into complex, risky projects with participants that are vital to our success, but we get them as cheap as we can.

I do not know how the Danish polar explorer Knud Rasmussen (1879 - 1939) selected his participants for the Thule expeditions in the early 1900'es – he writes nothing about it in his memoirs – but it was hardly lowest price. Five men and sixty dogs in the middle of the winter traveling from Thule into the unmapped wilderness of northern Greenland. Without sufficient fodder for either men nor dogs during the entire trip, but with the confidence that this would be found along the way in the form of musk oxen, reindeer, seals and other game animals, and with his own and the lives of others at stake.

I wonder if trust did have a high priority in his choice of participants, and not least trust in the participants' reliability when the unexpected happened, as it did again and again?

The same should probably apply to all other projects that, no matter how customary and routine they may seem, always imply a risk of the unexpected, that requires immediate action on the ground. Like when Sonja an hour ago suddenly discovered that the mayonnaise she was making, began to separate. I myself would have panicked and looked up in cookery books, but she just poured a little water into the bowl and continued. It is reliability and knowledge of how to handle critical situations without orders from above, that causes the system to operate.

It is this robustness the complex project expects. Well defined roles, clear, common criteria of success, understanding of the situation and the ability to act spontaneously on what is required. Last Planner!

Last Planner has become an expression of the management philosophy, which I was brought up with: Generate reliability in your organization. Make it simple, show confidence, share all information and delegate.

Your project participants can usually do much more than you think, and if you make them feel it is fun to participate, you're almost home.

As a young engineer, I had the responsibility for the expansion of the Greenlandic oil storages, including planning, engineering and supervision. On a monitoring tour to what was then called Jakobshavn and today Ilulissat I found that the pipes had been crossed in a pumping arrangement. They came nicely running along the mountain side, one above the other, gasoline at the top, kerosene in the middle and the fuel oil at the bottom. The pipes had not yet been painted, so it was a little hard to follow them when they dived through the filters and pumps, air separators and meters and came up again. However, crossed they were, so now the fuel oil was at the top and the gasoline at the bottom, which did not matter much for the function, except that we were quite strict with always having the pipes lying in the same order to avoid mistakes in the winter darkness.

I found the piping foreman and made him aware of the issue. It was big and strong fellow, he stared down at me and said that they were bloody not crossed and now he had been a pipefitter for twenty years and that kind of mistake he damn well did not make. Now I had met him the year before, so I just said quietly 'Twenty-one!'

Huh, he said. Well, last year you said twenty years, so now it must be twenty-one! The workers around him began to giggle, and I waited anxiously for the coming of a slap in my face. But then he split into a big grin, and when I then offered to bet a case of beer that I was right, we went down and looked at it. I still remember his face when he grabbed the gasoline pipe and followed it through the loops and suddenly saw where it would end …

The beers we drank jointly on the following Saturday evening and I was paying, but it created a piping team that the in the years to follow was extremely loyal and helpful.

The projects we undertook in those years were spread from Nanortalik in the south to Thule in the north, and they became a great success because we managed to create a joy of participation and being treated fairly.

And maybe also because our client, chief engineer Bøgekjær from the Greenland Technical Organization, had admonished me when I after a tender had pointed out a clerical error in the winning bid, where the price strictly to the tendering rules should be lower, but where Bøgekjær sub-sequently told me: You must understand engineer Bertelsen, we have to build oil storages year after year, and therefore we rely on our experienced contractors. If we don't treat them properly, when they make an excusable clerical error they may not want our projects, and it is we who have lost!

Wisdom, in spite of public money, strict rules and the state auditors in the background.

Cooperation and Self-Interest

We have recently passed the 70'th year anniversary of the end of the second world war and have not since then had an actual armed conflict between countries in Western Europe and North America, although we have been involved in too many wars elsewhere.

At the same time we have seen a historically unprecedented growth in prosperity and in Northern Europe also in welfare. For me who has experienced these 70 years, the peace and cooperation is the key to this success.

In the early 1990s NIRAS had quite a few projects in the former Eastern Europe, where I mainly occupied myself with projects in northern Russia. Here the former communist thinking that we're all equal and we do not cheat each other, still existed while everybody cheated to all of their capabilities.

I am not knowledgeable in this field, but I sense again a kind of Reynolds number. There are forces that cause us to hold back and cooperate, and there are forces that drive us forward to further our own interest.

The motto Liberty, Equality and Fraternity of the French revolution is a contradiction, said one of my colleagues, for liberty leads to competition, while fraternity leads to co-operation. And he was right, because as always, you need a balance between the driving and the braking forces.

I feel we here have the eternal triangle once again and think we will find the same balance in many social systems such as bees and ants, fish and birds, animals and indeed humans.

The Project is a System of Systems

Let me before I return to the project, just pass the ants and learn, as it says in the Old Testament. Many of us see an anthill as a collection of small annoying creatures that invade our jam jar, but in reality, their hill is a system of systems. Each ant may seem a simple insect, but together the flock possesses marvelous construction skills compared to their size. The termite may easily build hills up to 10 feet in height, and if we compare this to our own high of six feet, where the termite is maybe a tenth of an inch, we find that it for us humans would mean building houses up to a mile high. And doing it as part of their everyday work and quite fast. Looking at the indoor climate we again find marvelous engineering capabilities. The temperature is kept within a fraction of a centigrade, day and night just by 'ant ventilation'.

If we move closer, we find another system of systems. Ants – and perhaps especially termites – are depending as we know on wood. Partly to obtain material for their amazing building projects but also for food. But the individual termite cannot digest the cellulose that it finds in the wood pulp. Therefore it has in its stomach an intestinal flora of unicellular microbes, which degrade cellulose for the hungry ant. Now there is just the problem that those for the ant so useful microbes are not perfect either. For example, they cannot move round by themselves. Instead, they have each teamed up with approximately half a million bacteria which are arranged at their outer side with their tails flickering, and with 500,000 small engines they have a very good start.

These helpful microbes must of course also have energy, and to this end they team up with another kind of bacteria delivering the necessary energy against payment in the form of a constant supply of food from the ant's munching via its microbes in the stomach.

Indeed a system of systems on the forest floor.

The project is, just like the anthill, a systems' system – a system of living systems. The project is a living system, a kind of artificial life that exists along with other projects in the world of 'customers' who need the project's output as part of other projects, or productions in a complex interplay. Downwards the project is in itself a complex system, made up of agents and relationships with project participants as agents in the next layer, and then their departments, groups or gangs and individual employees... In a way, we talk of a mixture of natural and artificial life.

The British Tavistock Institute opened this universe for the first time scientifically in an anthropological study of construction site's cooperation in 1966.[3]

They did not quite reach to the bottom in this survey, but

they found however, that in everyday life, in principle there were five layers in the organization, but that it was only the top three that were shown in the project's organization chart. By contrast, it was the bottom two that in the daily life got the project done!

There is something here we do not yet fully understand, so let me take one step further.

The Social Logistics

My good friend and colleague Sigmund Aslesen from the Norwegian contractor Veidekke has introduced the concept of 'Social logistics' as a process creating harmony and coexistence in the project. Here also the concept of a common mindset – a common understanding – is introduced as something essential in the collaboration, which gradually grows during the project's life, if it is running well, but which we immediately tear away when the project is completed.

Sigmund and I came to know each other very well, when a dozen years ago we developed the Lean Shipbuilding concept for yards on the Western coast of Norway. Sigmund is a sociologist, but he caught immediately the lean project management idea, and together we achieved amazing productivity improvements in practice.

There is a whole industry that lives by helping to create such a common understanding. Basically, I am skeptical of its approach. Maybe I'm too old – or rather I am – but I have for a long time been in doubt when I met these people and their exercises. Others feel probably safer, but like falling in love is not something you learn, it is in the daily work that cooperation is created.

Sigmund is not such a hot air balloon, he is a practical person, soccer coach in his spare time and very committed to the club youth. He is also good to relax and to share a jar with after the long days at the yard, and not least he understands all the dialects of Norwegian spoken in that area, where I must apologize my own dialect, 'called Danish', as I always say.

Instead of making circle talks and other little school exercises Sigmund soon realized the idea, when I showed him the problem at the yards, and he helped the workers themselves to see the waste and suggest improvements.

Suddenly something happened that should happen. The collaboration became better, productivity grew and wastage fell, and after a few weeks the project to everyone's amazement suddenly was ahead of schedule, which was otherwise never heard of, they were always falling behind and the yard had to spend a lot of money expediting work to reach the date of submission. All in an overheated market, where subcontractor delays were a problem often turning up.

We only looked at the flow, and it was here we began. But later we found out that the earnings suddenly skyrocketed – money just poured in, because the waste time and therefore the costs fell. The extra costs we introduced were minimal, it was our own fees, payment for the process manager the yard appointed – he was there already but had previously worked with scheduling, but now he was repositioned – and a few costs for introductory meetings and physical incremental costs such as renting a construction site elevator for the vertical transport.

Everybody was amazed.

A Brief Digression to the Art of War

One profession we as project managers seldom look at is the art of war. War is – next to construction – probably the world's oldest project production if you dismiss creating and bringing up a family, and there are numerous studies of the nature of war and thinking about its management. The sentence: No plan, no matter how detailed lasts longer than the first meeting with the enemy, sounds familiar and it has been known in the art of war for centuries. Some attribute it to Confucius (551 - 471 BC), others say Bismarck's Field Marshal Helmuth Karl Bernhardt Graf von Moltke (1800 - 1891), the man who won the fatal war against Denmark in 1864 and who we Danes therefore remember, and again general

Dwight D. Eisenhower (1890 - 1969), albeit with saying Plans are nothing, planning is everything.

Today US General Stanley McCrystal together with some of his colleagues in 2015 published the book Team of Teams, that for me presents an overwhelming new understanding of the project and its management in a complex and dynamic world.[2]

McCrystal's background is his management of the US forces in Iraq after the Iraq War, where the battle was against the Iraqi branch of Al Qaeda. Here the world strongest army, trained in 20th century warfare, fought against a different army fighting a 21st century war – and was close to loose. During his command McCrystal changed the US thinking to the Power to the Edge principle, which delegates the operative decisions and gives everybody free access to all information available.

The first step towards such a common thinking – a common mindset – is in my own experience to talk nicely and learn to say sorry. Next to delegate as much as possible and trust the man on the spot. And finally to accept errors – and use them to learn, rather than to punish the sinner. The idea is to create a reliable cooperation without misunderstandings.

The latter is perhaps worth reflecting upon. A common understanding comes not by lecturing, but by working together and by synchronizing thinking in everyday life. Learning to know each other's facial expressions and gestures and getting to think in the same way, just as we see it in the best families.

Hal Macomber, an American management consultant, introduced me years ago to the method: Making and keeping reliable promises, that is, offer and maintain reliable commitments.

His idea was simple and logical, and I have used and attempted to use it again and again.

Basically the idea is that a commitment is more binding than an order, but that a commitment must be established through a dialogue between equal parties, and in collaboration. He also has a syntax for this dialogue, namely that I make a request, you make an offer, which I accept or reject. And then you report clearly back – either when the job is done, or if it can't be done as agreed. So:

- We must raise the scaffold half a floor, so the windows can be installed *(request)*
  - We can cope with that on Monday *(offer)*
  - Will you be completely finished Monday? *(clarification)*
  - Well, but surely Tuesday lunch *(precision)*
  - Then we have a deal. *(acceptance)*
  - And if you, carpenter, can start Tuesday at lunch time, when may you be finished?

Here the dialogue continues with the carpenter's offer, on when he can release the finished work to the next trade, and when the next, for example supervision, can make their inspection.

It is an agreement between equal men on the same level, and therefore not an order, and promises you keep if you possibly can as a skilled craftsman. It is part of our pride!

Some years ago, we were renovating the roofing of the 1970's apartment building where we live, and for many reasons it had been too late in the season. Planning and approvals had drawn out, so it was late autumn, when we finally got started.

Stupid from every point of view, but the roof was leaky, and water was dripping down into the apartments on the top floor, so it was a choice between plague and cholera. We had not called for tenders, but just asked our usual roofer, who had previously made repair work for his opinion, having a consulting engineer specializing in roofing, mold and stuff like that also sitting at the table. Together we found the technically best solution, and the roofer calculated a price, which our engineer checked and found fair. I added an extra sum for the contingencies I knew would come, the owners'

General Assembly said Yes – after some debate, Shouldn't we get an alternative price, it was said, but I said no, roofing and dental treatment is of the same nature, we want craft that lasts.

So we shook hands and started.

Along the way we had of course discussed the season, but the agreement implied that they should not be roofing when it was raining and the roof should be tight every day at the end of working hours.

The project started in mid-November and was addressed with a great eagerness. It was the beginning of the low season for the roofers and everybody was interested in this job, that would last well over Christmas and into February. A trailer was set up in the yard, but we had also established a coffee room in the hallway outside our own front door, where our 95 year-old neighbor Gerda took care of their morning coffee and a slice of her home-made cake, and the owners' association offered coke or soft drinks for lunch before Gerda took care of a cup of coffee in the afternoon.

I myself lived right next door, but as a matter of principle, did not interfere in the process, which our engineer super-vised. But then, all the same we were there all the time and the craftsmen knew very well they were 'building for Gerda, Sven and Sonja'.

The unforeseen problems came of course almost imme-diately from the start, but here I was as a client ready with money in the budget, and were the extra cost excusable they were paid immediately. And believe me or not: They came, but only those that were justified, so we did not talk much about money, but concentrated ourselves on the process itself.

It was agreed in principle that they should remove the harassed the old and leaky lining section by section, all the way down to the concrete structure and then smooth out depressions with liquid asphalt, and finally put on the first layer of roofing felt giving us a water tight roof again at the end of the day. A fine rhythm, but in their eagerness they started a little faster the first morning and opened too much roof.

Oops! In the afternoon they discovered that they could not make it. Then there was sent home a message to get flood lights and extra heaters sent to the site, pizzas were bought – and they humped on. Along eleven p.m. they closed down and went home tired, but the roof was tight, and the 'client' – myself – who had followed the course with excitement, was happy.

Halfway in the process we as owner invited everyone to a 'topping-out' at the near-by Cafe Emil and used some of the reserve, because you may well show appreciation along the way, if you are satisfied with the work being done.

We came to like our craftsmen, and I sensed that for them this was a nice site. And actually, I also think that we got a better quality.

The schedule was kept, and the roof is tight. Our heating bill has fallen and everyone is happy. We gambled on good cooperation, and we won.

We are not together in the project to cheat each other, if we take the concept of project management seriously. And my experience is that such informal agreement form, naive as it may sound, often leads to better value because it creates cooperation in the project's everyday life, just as Last Planner suggests.

Of course, over the years there have been some who have tried to exploit my naivety. But there have been very few in my long life in construction. As the roofer said at the hand over lunch – his invitation: If all projects were like this, many leaky roofs would be avoided.

One should never buy the cheapest, but always the best, for the best is always the cheaper in the longer run, as we said in NIRAS with small poem – a so called Gruk – in Danish which the Danish poet Piet Hein wrote to us.

In my translation it goes:

BEST ADVICE

Good advice can be expensive,
Says an adage, fools believe
Ah, such old and empty words
  Do you often trust the deepest.
Choose the wiser substitute,
Change the older words with newer.
Cheap advice can be expensive.
More worthwhile is good advice – but
  Best advice is always cheapest.

1) Alberts, David S. and Hayes, Richard E: (2003): Power to the edge, command ... control ... in the Information Age. DoD Command and Control research Program, Washington, DC

2) Mc Crystal, Stanley; Collins, Tantum; Silverman, David; and Fusell, Chris (2015): Team of Teams: New Rules of Engagement for a Complex World. Portofolio/Penguin

3) Tavistock Institute (1966): Independence and Uncertainty – A study of the Building Industry, Tavistock Publications, London

The living project

*Where I recognize that shit happens,
and realize that projects may be managed
– but people must be led*

IN 1995 SONJA AND I TRAVELED around the world looking at the building process elsewhere. The tour became an eye opener. We met the most exciting industry people in the countries we visited, and we were everywhere received with open arms and great interest in our own studies.

I tell you that this kind of travel can be a renewal of the whole rest of your life. We were at that time at the end of our fifties and still receptive to new ideas, and they poured forth. One of them was partnering, which we first heard about in the United States, but even more in Australia.

The concept sounded excellent and seemed to be the solution we were looking for, and back in Denmark I brought it forward in the debate we also then carried out about the building sector's low productivity. The idea was taken as the quick and easy solution and no one took the time to think it through. Neither did our Australian colleagues, I learned when we met them again a few years later.

It's one of those ideas that – as Winnie the Pooh says – sounds good in your head, but does not work outside it, while he hangs in Christopher Robin's balloon outside the bee cavern with its honey.

My understanding today is that there are no simple solutions in the complex project, where cooperation is the key to better productivity, but everything else resists. Cooperation works, and even infatuation, which in the adventure can overcome all, it is not allowed in the project. Lowest price sounds a battle cry, and lowest price it becomes. Even if there were other companies we much prefer to participate in the project.

That the sum of the lowest prices never gives us the lowest costs – almost certainly the contrary – we should have realized by now. Lowest price leads to sub-optimization and bottlenecks, while the flow becomes unreliable, everyone loses and claims rise increasingly, which in turn destroys the cooperation, while the economy runs out of hand. A US study indicates that contractors base their calculations on a flow reliability – a PPC – at 50%. It is expected, therefore, that every second task in the weekly plan will not be executed as planned – which very precisely corresponds to what can be seen in practice – so it's probably true. This beats the unreliability of the train services by several lengths and brings many projects far below the precision of the met office's weather forecasts.

The solution is perhaps better collaboration and cooperation, which also partnering suggests, but cooperation must be built bottom up, and by showing trust and entrusting responsibility to those with the hard hats on.

Systems may be Controlled but Humans must be Lead
The more I wonder about this new understanding of the project and the new approach to project management, the more I recognize that it might not be a matter of controlling

the project, but to lead the participants in the project's realization.

Quite a few years ago my good friend and colleague Sigmund Aslesen called it Social Logistics in his sounding Norwegian. A really good description of the term which I have also met named Common Mindset or Common Understanding.

The British management consultant Alan Mossman, when we discussed the concept, called it the eighth flow, but it is not so in my understanding. I see it as a common understanding created through a social process that supports the handling of the project flow. Unity and mutual understanding must grow by itself just as love, we must care for one another and experience the joy, of creating something of meaning through the project.

I would rather say that social logistics is a side product in the good building process. It should be fun for everyone to participate and to work with the unruly project!

For that to happen, the project in my experience must build its own culture, in the same manner as is done in good companies, associations – and especially families. There must occur common understanding, language and behavior that support collaboration and unity, even when chaos threatens and everything is burning while critics are poised.

Building a Common Mindset is an important task in its own right.

But do we do that?

In my view, we do just the contrary and we do almost everything we can to hinder this approach. Our entire traditional process of selecting the project participants is based on distrust. This is reflected clearly by the often hundreds of pages of contracts; in the minutes of meetings construed as legal documents with the tribunal in mind; and the participants' search for loopholes in the contractual basis, so they can offer a low price to win the contract, and then later milk the client through these holes.

As a client, we certainly buy many of our problems ourselves.

This is by no means a Danish phenomenon. In the claim-oriented practice in the United States the situation is even worse. Here the building process is almost choked in forms that take all the time of the middle managers who should ensure the productivity. Because one can be taken to court with outrageous claims, if an odd form has not been completed in time, paper-work is constantly superseding solving actual problems. I am feeling that we in Denmark may be heading down the same slippery slope. Each bidder – and especially those who win – spends quite a lot of time seeking the largest gaps in the contract basis instead of finding the best process.

And this is happening at all levels, from the overall contracts to the appointment of the latest Bulgarian laborer. Lowest Price sounds a battle cry out over the construction site, where few dare stop and think about whether there might be a Best Price?

If we forget our financial department's spreadsheets and our lawyers' sophistry and instead bring flow into the picture, we will be looking at a quite different situation. Can we create cooperation and reliable flow, plenty of gains are waiting. Shorter construction time, fewer errors, increased safety and yes, believe it or not, a far better economy.

All Must be Onboard

To do this, we need commitment right from the outermost part of the organization, interest and empathy in the project's progress, and it must be reached by giving responsibility and encouraging improvements and creating ownership and cooperation. Not without reason, lean construction has been called partnering from the bottom, where partnering otherwise only takes place in the upper echelons of the organization. Last Planner provides for such a cooperation, and it is typical that the workers immediately take it to their heart when I introduce the principle. The same was also my experience when I presented the ideas one morning fifteen years ago at a meeting with the Danish Building Workers' Union. The thoughts went directly home, and they were immediately joining the bandwagon, something several of my foreign colleagues have wondered about, because to them the unions are always an opponent.

But for us in in Denmark these new principles are just fitting into the very successful Danish model of trade agreement. Last Planner provides not only a better flow, it also generates job satisfaction, a better working environment and better piece rates in the trades, where they still use them. And then there is also the matter of the enhanced safety.

Safety? Unfortunately, there are few data, but those we have show that the rate of accidents drops dramatically, and is often almost halved without doing anything other than ensuring reliability in the flow, and only allowing tasks that crews themselves consider sound to move ahead and get into the weekly work plan. Then it just happens, and the better we understand each other, the more often it happens. With the right to say No we are namely handing over a big responsibility, and in my experience this is received with similar seriousness. The few results we have, are cited quite frequently in scientific papers, they are something that gets noticed.

Over the years, I have learned that most crews love to work independently, and that my problem as a leader rather has been keeping too ambitious initiatives under control – in the same way as my colleagues over the years have done with me, when I went into overdrive. But complex dynamic systems are moving by themselves towards their critical state on the edge of chaos where they operate optimally, which indeed is what we want them to do.

As my riding instructor said when the horse had to be kept in step and held ready for the transition to trot or gallop: you have to push forward and hold back! This is how I experience the good project as well. It must quiver with energy, ideas and initiative, and as a project manager I should carefully have to hold back, but at the same time stimulate further initiative.

In most cases this Last Planner approaches works: Build trust and create reliability, delegate and give responsibility and accept that errors occur, if we want to work on the edge of chaos.

Shit Happens

In our Christian faith, we are brought up to consider errors a sin and that everything we set out to do, also has to succeed. I remember the mental shock I experienced when a good friend and scout leader told me about his master thesis in electrical engineering, where he was investigating whether a new method for the measurement of the electrical resistance could be used in practice, and how he had found that it could not. I saw it as a failure, but he pointed out that if trials did not from time to time fail, they were not trials, and that you actually learned more from your mistakes than from your successes.

That day I learned a lot.

Later I found that this wisdom is one of the secrets behind Toyota's production system: They treat errors differently from us in the West. Our finance department downstairs on the first floor considers the error as a cost simply and notes it on the red side of the accounting. We who erred are ashamed and try to forget it and sweep it under the carpet, as a Danish prime minister once said.

But how much do we lose by this attitude? If instead we saw our mistakes as assets – as investments in a better process and as necessary trials – we would get a capital of knowledge and experience to draw on in our further development. Is there an error, we should stop and find its root cause. We should then remove it, so that the error never comes again. Only in this way, can we hope to become flawless, as Shigeo Shingo argues.

It is the same idea behind PPC in Last Planner, namely

the search for reliability, not for sinners, and together becoming better.

Mistakes happen, that we must accept. Just as we must accept that unlikely events also happen in our project. The Improbability principle applies everywhere in complex dynamic systems. The unlikely will always happen, it's almost a law of nature.

Let us therefore use our mistakes to learn – and for all the world not be tempted to build more rules and control systems to avoid them. Errors are like weeds, they will always find a way around our obstacles, and it is only through the man on the floor we can fight errors, and then use Shingo's advice to find and eliminate the cause, so the error does not recur.

Which it nevertheless does in the project's complex world, especially when the flow slips into the turbulent – the chaotic – state.

Or if outsiders start to interfere.

Keep your Hands Away

When something does not seem to work like it should in the delicate system of the project, when it is teetering on the edge of chaos, it needs little baffled interference, before it ends up in total chaos.

Therefore, my advice is to stop, breathe and quietly find the cause of the critical situation. Not to spend a long time, but to think before acting, and take the time to listen to the man on the site.

A good approach is often to find the critical flow, i.e. the flow which contains the source of the problems. With this aim one has a logical approach to unravel the problem, to follow that particular flow backwards and to find the bottleneck that is likely to be the cause. Then it is necessary 'just' to ease the bottleneck, usually by highlighting it and get the participants themselves to do something about it. In other words, help the system itself find its way out of the crisis.

Top-down managing of an anthill is never going to work.

Strengthen the Culture and Create Engagement

But what is the client's role, if he should not interfere when things seem to slide?

It first and foremost is to provide a framework for things to happen and the project to unfold – and then to stimulate the project's participants to engage themselves in the project.

It sounds beautiful, many people will say, but can it be done in the real world? It may, for I've seen it again and again in recent years in Denmark. The process of constructing the sculptor Bjørn Nørgaard's architectonic beautiful housing at Bispebjerg Bakke is one example, the Housing Association Fruehøjgaard in Herning is another and fortunately the list grows all the time, so there may be light ahead, although there is a long way home.

Go and See

In the Toyota concept – which I have avoided until now, because we do not make cars but projects, and because I only know it through the western interpretations – one part is Gemba – the spot where things take place. In our Western interpretation it means Go out and See where the work is happening, and it's an advice I give again and again when I explain my thoughts. Just go out and see what happens in your own production. Move your eyes away from the work creating value, to what should not happen, to the waste. Surprisingly observations will topple forward, a true self-service buffet for an improvement process.

I recently participated as coach in a large project. Our process planning included multiple-day meetings, and each day the lunch break was marked by the long queue at the buffet. Everyone waited patiently, and then I asked the question: What does not work here? Why do we not just get our food and start eating?

There were several proposals which we discussed with food in our mouth, and the concept of flow began to be understood. The next day, I got the buffet moved 3 feet away from the wall, and hello! Now the hungry participants streamed down along the same buffet table, but with double capacity.

The queue from the previous days melted away like dew in the sun, and the participants saw that to increase the capacity of a flow system is often surprisingly cheap and easy if you look for congestions.

Frankenstein's Monster

I have earlier several times talked about the project as a complex system, and pointed to complexity theory as a new start. It is a theory which, not least as it unfolds in these years at the Santa Fe Institute in New Mexico, is incredibly inspiring and often provocative. So, what about trying to see the unruly project as artificial life?

Just try the idea. My thinking is that the unruly project is a man made, living being – as Frankenstein's monster – something we ourselves have created and set to work, and which is now running away from us. If we once again abandon our rational thinking, relax, take a step backwards and look at the monster we have created in the form of our project, then the project may in fact look similar to a living system. Not only as a living system like the tree in the garden, but as a real living system.

Many of the computer games we encounter these days are artificial life. The games have their own ability to develop, learn and exhibit new and surprising patterns.

The pharmaceutical industry uses the same ideas in their research in the effects of new drugs. The financial world, which always has big money available and is quick to respond to new ideas, has also seen the theme as interesting, as its massive commitment to the Santa Fe Institute testifies, so there is probably some truth to the idea.

In my sixth essay on The Autonomous Project, I spoke about the project as a system of systems, linked to other projects in an infinite and eternal network of systems, usually with multiple systems in each layer, where they interact. At the same time a development of this symbiosis takes place, which in many ways resembles a living system, and the question is whether this is how we should understand the project.

When I have used the word unruly, it is also because I see the project as something alive in its own right. Not the life each participant adds, but the project itself as something alive.

This living organism has a function in a greater cooperation of similar artificial organisms, which is to create something. When this something is created, it dies, and its parts are so to say eaten by other projects. The organism itself is complex and dynamic. Not only in its function, but also in its ability to renew and incorporate new systems, while others, which have no more tasks to perform, are repelled.

We have in other words an ordered world of production as our basis, and a chaotic world of innovation and renewal in the form of a flow of new projects alive and kicking.

Artificial Life

Artificial life is a science that today is flourishing in many places all around us. In short, it is an approach to the study of living systems through computer simulations. The first attempts were probably the English mathematician John Conway's Game of Life from the 1970s.

He created in his computer a single two-dimensional pattern of live and dead cells in a grid, and then let the system develop generation after generation from two simple rules: A living cell survives if it has two or three living neighbors; and a dead cell wakes up if it has exactly three living neighbors. Neither more nor less, and a whole universe unfolded itself on the screen. The program is trivial to write, we wrote and executed it ourselves on the old VIC 20, and depending on the starting situation the most amazing patterns turn up.

Prior to the first scientific conference on artificial life at Los Alamos in New Mexico in 1987, Craig Reynolds presented something more sophisticated, an algorithm that simulated a bunch of animated penguins in the movie Batman Returns. Once again quite simple rules, but it worked. He presented to the conference the program Boids which he had developed, that simulates a flock of birds and again by two very simple rules: Follow the others and avoid flying into them or other obstacles.

Later, work has continued with similar systems, where quite simple rules govern the interaction between the individual, artificial subjects, and with amazing results. An obvious way to study the project – for is the project not at the end of the day a man-made life?

Talk Nicely to Your Project

After one of my courses in Lean Construction a participant came up to me and said that what I had told reminded her of some 'important words' that they used in her company. It was quotes from Peter Kiewit, a British engineer and owner of Kiewit Construction, and they were:

The least important word is **I**
The most important word is **We**
The two most important words are **Thank you**
The three most important words are **If you please**
The four most important words are **What is your opinion?**
The five most important words are **You did a great job!**
The six most important words are **I admit I made a mistake.**

I think that his 22 words is the way to a better management of the unruly project.

A New Science

This concludes my speculations on the Unruly project, but it by no means solves our problems with the phenomenon. The vast majority of our production – after industrial production has been outsourced to the East, or has been taken over by robots – is presumably happening as project production. I have not been able to find precise figures from Denmark on this, but it is my firm conviction.

In this light, the project is a strangely neglected kind of production, which costs us dearly. I have in these seven essays tried to show how we are wasting a lot of money and missing a lot of value, because of our poor understanding of the project and therefore our primitive approach to its management.

Not that we lack guidance in project management. We have abundant textbooks, courses and consultancy offers. But we lack a deeper understanding of the nature of the project. Project management, as we today teach it at the universities, is a subject that floats in the air without a proper theory, said an Australian professor to me some years ago.

This may, in fact, be the problem. For in the absence of a deeper understanding based on solid theory, the project remains for most people simply a series of operations, where the challenge just is getting them executed in the right order and at the lowest cost.

An important step on the road to a new project under-standing is therefore in my eyes, that we recognize the study of the project and project production as a new science, which derives its theory from a number of other sciences.

This leads directly to my next consideration, namely who should take on the task of establishing such a new science. Initially it would be tempting to think of the institutions already involved in project management, for example in relation to building engineering or the management schools.

None of these, however, have even caught sight of the

need – and at least not taken initiatives in this direction. And it is also a question whether such a new science should be established free of a particular traditional science or industry-related angle. Project science will have at least as much relevance to the many other sectors in our society and business, that is making use of project production – the IT world, pharmaceuticals, event organizers, movie producers, fashion designers and many more.

But when we in construction are one of the players with the longest project experience, it is natural that it is us who kick-off. Maybe we should try a totally different approach, where we gather inspiration from the Santa Fe Institute, which – as I have pointed out several times – despite its modest size in a few years has built up a professionally strong and inspiring environment in the study of complex systems, which has gained international clout. It has especially happened by bringing leading scientists from different disciplines – often Nobel Laurates – together in a creative environment, in the early years in an old monastery, a sort of refuge where scientists from all over the United States came and stayed a while, worked, discussed, went walking in the desert just outside, occasionally drank a beer or two, and eliminated many of the so called two beer questions.

When I read about it, it reminded me very much of my feeling of what happened at Niels Bohr's institute in Copenhagen at the beginning of 19'hundreds.

Despite the modest surroundings they were thinking great thoughts.

I am not in any way that ambitious. But I experienced myself something similar albeit in a more mundane scale when the Danish Association of Consulting Engineers in the 1990s under the leadership of the very engaging director Tage Dræbye (1943-2013) established groups of members to discuss subjects of general importance to society. For a number of years, the association set an agenda for the development of the building industry, and many – members and non-members – participated with great enthusiasm.

My vision is that we in Denmark establish a similar multidisciplinary international institute to develop a new project understanding through a cross-border cooperation – and through development and tests provide new insights and new tools to further enhance the productivity of our projects. Not only in construction, but far into other industries from research to theater production, from IT systems to product development.

Not an academic institute which by itself has to conduct research but a center which in a qualified manner could bring practitioners and researchers together in brand new mixes. Perhaps rather a proactive and dynamic, creative think tank that through publications, lectures, workshops, seminars and conferences boost a mental shift in society's many project producing branches.

Acknowledgements

This book has been many years in the making and there are many that, directly or indirectly, have contributed to these seven essays. This lies in the process' nature, and often my thoughts occurred during lectures, conversations with colleagues and not least students, in the deep armchairs in my office in Holte, an untraditional workspace furnished one hundred percent with an aim to reflection, not to actual meetings.

Many of my thoughts have then been scribbled on the back of an old print, and later – often late at night – translated into a Trimmed Thought – one of my Friday blogs later to be published or just put in the pile with others thoughts, waiting for a book like this.

I highly appreciate these conversations, which nearly every time has given me new questions to think about, and I would be sorry to do without them.

However, there are others I have had more stable conversations with, especially the three Musketeers, my good personal friends: Greg Howell, Glenn Ballard and Lauri Koskela. We four meet every Tuesday afternoon for a chat on Skype, which is an amazing inspiration for me. And not least thanks to Glenn Ballard in particular for patiently following me along the very long journey in sorting out my thoughts and getting them down on paper.

There are also many excellent colleagues, I have met in my professional work that have inspired me and still trigger new ideas. Not least the recent construction projects at the Danish Technical University campus was a source of much learning, with both positive experiences and with disappointments from time to time. But I hope to see DTU as both a college and a highly professional client that can play a role in the Danish construction industry by developing new understanding of the project and immediately test the ideas in full scale. I have seen Sutter Health in California doing this over a dozen years through a construction and development of the same magnitude as the ongoing program for super hospitals in Denmark.

The association Lean Construction – DK received me nicely when I asked them for help in the final editing and publishing of the Danish version. Randi Muff Christiansen, the then energetic chairman both commented on my manuscript and found the necessary financing from the Realdania foundation. It brought my Danish editor Poul Høgh Østergaard into the process, for which I am very grateful. Indeed, it was a quantum leap for the book. Poul managed in a series of very long interviews in the deep chairs over many Monday afternoons to get me to talk, while clarifying my thoughts and finally writing the book all over. And with Poul came also Claus Lynggaard that in a highly professional manner turned my writing into a book.

Thanks to all of you.

And finally, there is Sonja as always. When I published my previous book Semiramis, talking about project physics I promised her that I would write no more books. But these essays were not a book when I started the project. They were my notes from a series of talks, lectures and discussions I would write as a small hand out compendium, but along the way I found what so often happens in complex systems, something unexpected arises. Emergence as it is called in complexity theory.

And thankfully Sonja, whom I have been in love with for half a century, accepted this development and has since patiently commented on many of my loosely written thoughts. She has also accepted that one holiday after another was spent sitting at the computer and writing, instead of normal holiday activities.

I owe in other words, a big thank you to the many that have helped to get my thoughts out to the world. Now I just hope others will welcome them and maybe find inspiration to carry the ideas on. For I have far from reached the end.

Holte, November 2015
Sven

SVEN BERTELSEN

Born 1937 and MSc in Civil Engineering 1961, Sven Bertelsen has spent his whole professional life among projects. For forty years as a consulting engineer in NIRAS – a leading Danish consulting engineering firm, where he for twenty five years was a senior partner, in charge of a number of NIRAS' major projects, including NIRAS' key role in the successful establishment of the Danish natural gas supply system.

Today his name is first and foremost connected to his work within the International Group for Lean Construction – IGLC – where he since the late 1980es has contributed substantially to the development of the theory, where not least by his ideas about understanding the project as a complex, dynamic and adaptive system are internationally recognized.

He introduced lean construction in Denmark and was a co-founder of the association Lean Construction-DK, of which he in 2014 became its first honorary member.